Y0-CDK-471

Mechanics of Impression Evidence

Mechanics of Impression Evidence

David S. Pierce

CRC Press is an imprint of the
Taylor & Francis Group, an **informa** business

Auerbach Publications
Taylor & Francis Group
6000 Broken Sound Parkway NW, Suite 300
Boca Raton, FL 33487-2742

Printed in the United States of America on acid-free paper
10 9 8 7 6 5 4 3 2 1

International Standard Book Number: 978-1-4398-1370-6 (Hardback)

Visit the Taylor & Francis Web site at
http://www.taylorandfrancis.com

and the Auerbach Web site at
http://www.auerbach-publications.com

Table of Contents

Preface

Fingerprints were a popular topic of the day in the late nineteenth century. It is entirely possible (as proposed in some circles) that Samuel Langhorne Clemens was influenced by the treatise *Finger Prints*[1] when he wrote the following[2]:

> I beg the indulgence of the court while I make a few remarks in explanation of some evidence I am about to introduce, and which I shall presently ask to be allowed to verify under oath on the witness stand. Every human being carries with him from his cradle to his grave certain physical marks which do not change their character, and by which he can always be identified—and that without shade of doubt or question. These marks are his signature, his physiological autograph so to speak, and this autograph cannot be counterfeited, nor can he disguise it or hide it away, nor can it become illegible by the wear and the mutations of time. (p. 134)

Mark Twain is not included in any history as an expert in the science of identification practices. His works of fiction were, however, able to grasp public attention and elicit acceptance far easier than any contemporary scholastic study. His vision has been tarnished little by the strain of over one hundred years of scrutiny.

Today, we are faced with increasingly complex forgeries, blatant errors, and challenges in the courts; yet, the underlying truth in those homespun phrases still rings out. It is a truth evident in, and applicable to, all forms of impression evidence.

Forensic science is composed of varied disciplines that must work together to achieve a desired result. The bond between laboratory and fieldwork is separated by methodology and scope while connected by the common goal of proving facts.

Not so long ago, the functions of footwear, tool mark, and fingerprint examination were duties carried out by the same practitioner. "Identification experts" often took pride in the diversity of areas in which they could claim expertise. These were the generalists.

With the growth of varied technologies, it was logical for specialization to occur. In many respects (particularly from an administrative point of view), development of specialist areas is an efficient way of coping with the matching of a worker's abilities, interests, or capabilities with a burgeoning wealth of new information and required training. In short, specialization is cost effective.

There are a few locations, globally, that recognize the importance of a well-rounded forensic education. In those locales, the training includes exposure to each of the specialized branches of the craft, thereby ensuring a proper fit for a well-rounded employee.

In 1937, an author recognized that there were three important groups of scientists and engineers who shared a common interest in elasticity. Physicists expressed their observations with formulas. Chemists carried on the work in the laboratory, and engineers often took the raw information into the realm of applications. Houwink and de Decker's book, *Elasticity, Plasticity and the Structure of Matter*,[3] offered some common ground for the varied disciplines to share insights, improve existing materials, and discover new ones. The initial task, linking common observations about elasticity and plasticity, was found, by him, to need the addition of an introduction to the structure of matter.

This text pays homage to Houwink's insight by offering a similar attempt to provide a look at the commonalities of evidence as well as some particular obstacles faced by related forensic disciplines.

A gentle warning for those who follow the phrase "best practice" as adopted literally by some and practically by others. Best practice can only exist in the moment and can be expected to change. Adherence to what one believes is best practice may deter some practitioners from extending their abilities and others from "raising the bar." This text, therefore, is presented as a guide and a form of encouragement to those who seek to answer the questions posed by every scene of crime.

References

1. Galton, F. 1892. *Finger prints.* Macmillan, New York.
2. Clemens, S. L. (Mark Twain). 1897. *Pudd' nhead Wilson.* Common Place/Simon and Schuster, New York.
3. Houwink, R., and H. K. de Decker. 1971. *Elasticity, plasticity and the structure of matter.* Cambridge University Press, Cambridge, UK.

Acknowledgments

We tend to produce an amalgam of that which molds us. I have been very fortunate to have met and been influenced by a wide group of diverse people and ideas. Teachers aspire to inspire and there is no shortage of good teachers in my past, from Professor David Korff, M. F.A., formerly of Lambton College, to Harold Tuthill, retired forensic instructor at Ontario Police College, Aylmer. Good friend, Shane Turnidge, rekindled my interest in forensic research. Professor Bernd Kaiser, Lambton College, willingly donated time and effort with advice that was pivotal, given the short span of time in which research, experimentation, and writing had to be completed. I am grateful to the University of Western Ontario, Research Park, and managing director, Don Hewson, PhD, for assistance and support, and the graduate student, Patrick Mallay for his efforts.

A special thanks must go to contributors; Sandra Wilson, a family friend, who earned her PhD shortly after contributing the first chapter; Eugene Liscio, who is on the right track to contribute much more precision to the world of forensics; and Bill Lee, PhD, who (although he didn't write for this book) gave permission to include some of his findings regarding electrostatics.

Acknowledging family support is a much different matter. My wife, Sally, offered needed criticism and many hours of her time. My wife's aunt and uncle, Rita and Ron Baker, have always provided inspiration and support, as have her father, Fred and her mother, Mary, who will purchase the first copy. Two of my sons, Geoff and Phil, diesel locomotive mechanics, gave up their boots and a little technical support for the research, and my brother-in-law, Christopher Adcock, PhD, provided advice. This was definitely a group effort—thank you all.

The Authors

David S. Pierce is a graduate of the Ontario College of Art (OCA) in Canada. While attending OCA, he learned about, among other things, injection molding machines, casting techniques, industrial design, print making, and the general arts.

During postcollege employment, he learned to operate printing and industrial presses and worked with graphic design. David also worked at Lambton College, Sarnia, Ontario, Canada, both as instructor for the radio, television, and journalism programs and as support staff. In January 1983, he began his career with the Sarnia Police Department, and in March 1984, he completed his training as a forensic investigation technician. David has received extended training at Ontario Police College, Aylmer, and privately pursued further education in areas such as footwear and tire track training. He is certified as a footwear examiner through the Canadian Identification Society and is a candidate for identification recertification. David has had three articles published by the *Journal of Forensic Science* and another by the Canadian Identification Society. He gave a lecture at the 2008 International Association of Identification Annual Conference in Louisville, Kentucky, and made two poster presentations at the same event. Perhaps more important, for this book, he invented several devices, techniques, and equipment that can only be termed as thinking "outside the box."

Patrick Mallay is an undergraduate student at the University of Western Ontario. Patrick spent the summer of 2009 working at the prestigious University of Western Ontario Research Park in Sarnia, Ontario, Canada. Under the auspices of the director, Don Hewson, he created the basic foundation of research presented in Chapter 5.

Patrick found the work interesting, and his assistance was instrumental in turning the input of the author into a useful and informative study.

Eugene Liscio is a registered professional engineer in the Province of Ontario, Canada, and is a graduate of the aerospace engineering program at Ryerson Polytechnic University (Toronto). Eugene started his career in the Materials and Process Engineering Laboratory for Boeing Canada (formerly McDonnell Douglas Canada Ltd.), where he was responsible for the review of metrology tests and conducted many failure investigations. Eugene then moved on to work for a U.S.-based aerospace company, where he used his

3D (three-dimensional) design skills to create concept machinery. At this time he became interested in how 3D tools could effectively communicate ideas and concepts; he then began actively pursuing work in the areas of legal graphics and animations. He eventually began to focus on crime and accident reconstructions; however, today much of his work is to provide forensic mapping services.

Eugene Liscio is a member of the IAFSM (International Association of Forensic and Security Metrology) and the ASPRS (American Society of Photogrammetry and Remote Sensing) and is actively engaged in research utilizing 3D technologies in the application of crime and accident events.

Sandra Wilson earned a bachelor's degree in microbiology from the University of Calgary, Alberta, Canada. She is currently a doctoral candidate at Queen's University, Ontario, Canada. Her current research focus is microbial ecology with an emphasis on molecular techniques. While not trained as a botanist, she finds the molecular intrique and challenge inherent in forensic botany as well as the ethical ramifications appealing. This is a burgeoning and promising field; however, the molecular techniques and forensic application must increase in concert for this field to truly flourish.

Introduction

This is a small book about big topics that applies to the larger concern of evidence. The topics are mostly new to forensics even though the principles are common enough in scientific circles. The subject matter consists of unique observations regarding specific circumstances that are intended to generate more questions than they answer.

The style of writing was chosen with serious practitioners, students, members of the legal profession, and laypeople in mind. While the content does not shy away from the science that is the future of forensic practice, the topics are mostly simple. The purpose of this approach first is to share observations and discoveries and second, perhaps more important, to inspire practitioners and students to look beyond what is known to perform intuition-based research; finally, the book encourages the sharing of findings that is hoped will generate a more unified approach to the topic.

There is an unfortunate tendency to segregate one forensic discipline from another. This is reflected in training, literature, and practice, all of which can benefit from acknowledgment of similar principles and (in many cases, such as localized environmental effects) the same formative issues. Segregation seems particularly outdated in light of current trends toward interdisciplinary activities.

As an example, two forensic experts, one trained in friction ridge identification and another in tire track analysis, work the same scene and never compare notes about the nature or circumstances of the impressions they find. It happens. It is analogous to studying a subject like mathematics, for which a grade of 80% on an exam sounds good without considering that it can mean that the individual missed 20% of the subject matter. A problem could develop, over time, as there will eventually be a need for that missing knowledge.

The estrangement of one forensic discipline from another is escalated by language. In the general sciences, the language used is global. The techniques, measurements, conclusions, or methods for one scientific endeavor do not differ from another in form. This global approach seems a worthwhile objective for the forensic disciplines.

Studies of mechanical properties are "those responses a material has to the application of mechanical forces."[1] Typical responses of materials include deformations resulting from stress and strain. An impression results from an interaction of more than one material, and the various interactions that do occur are the "mechanics" discussed in this text.

Change is both needed and appropriate within all forensic disciplines because without change there can be no growth. Future change to forensic concerns will prove the usefulness of the scientific method to all forensic disciplines. The content of this text is also intended to assist the transition toward scientifically defensible forensic practices.

There is no pretense in this book regarding the complexity, number, or variety of cases solved by the techniques and suggestions that are offered. The answer, if such a question were posed, is that few known cases at the time of writing have made anything but tertiary use of the concepts presented here. Much of past forensic research and literature accumulated over the years stems directly from casework, whereas the material presented in this book is essentially new ground.

Some of the suggestions and exploration included in this text may point the way to further research without ever attaining popularity as "accepted practice." The point here is not to dictate which procedures to use but rather to allow practitioners the benefit of alternatives that may one day mature into accepted techniques. The difference between success and failure in investigations can often be traced to issues that restrict our actions to those that are familiar or anticipated.

Casework can provide a valuable training tool, but research is the next level of education. Research can be instigated by an observation or an inspiration and needs to be unencumbered by the demands of a particular case. In support of research, it is wise to seek the affirmation of the scientific community for both findings and questions of methodology.

Practitioners are seldom serious researchers; even though they are uniquely positioned to make use of their observations, they generally lack the time and means to follow up on the intuitions. The value of a more innovative approach is likely to increase without the stress of having to deal with the immediacy of challenges presented by a particular scene of crime. In other words, there must be a separate time for research even if that time is devoted to a particular case.

Remember that information concerning impression evidence should answer the same five questions that should be answered in any form of communication: Who? What? Where? When? Why? In the practical world, there are times when impressions will answer those questions more or less thoroughly and reliably than DNA evidence. This book assists in understanding the principles and situations that can often answer a sixth question: How?

In 2008, a report by the American National Academy of Science[2] recommended an overhaul of forensic science methods aimed at developing scientific and accountable methods in almost all areas. While advocating wider applications of the scientific method, this text proposes that, at least potentially, both deductive reasoning and empirical methods are correct and applicable.

References

1. Van Vlack, L. H. 1973. *A textbook of materials technology*. Addison-Wesley, Reading, MA, p. 9.
2. Report of the National Research Council of the National Academies. 2008. *Strengthening forensic science in the United States: A path forward*, prepublication copy. National Academies Press, Washington, DC, http://www.ndsu.edu/dna/documents/ NAS%20report%20overview.pdf (accessed June 20, 2010).

Forensic Analysis of Wood DNA

1

SANDRA L. WILSON
DAVID S. PIERCE

The first instance in which data from plant DNA was accepted as admissible evidence in a criminal case was in Arizona in 1992. This was a case that placed a suspect's truck at the scene of a crime by virtue of the evidence gained by comparison of the seed pods in the truck to those of a Palo Verde tree (*Parkinsonia microphylla* Torr.) at the scene. Dr. Tim Helentjaris, a geneticist at the University of Arizona, agreed to try, using randomly amplified polymorphic DNA (RAPD), to produce profiles of visualized DNA fragments.[1] This case offers an example of the appropriate specimen collection procedure for comparison by Maricopa County investigator Charles Norton. The idea behind this chapter is that wood fiber is too often neglected or ignored as evidence pertaining to a suspect who has forced an entry during the commission of an offense. The act of prying a door or window (commonly made of pine), for instance, can result in the creation of fiber evidence that is too small for a physical match to the pry marks. Those fine fibers logically may yield useful DNA evidence capable of placing the suspect or the suspect's clothing or tools at the scene. This first chapter is set apart by the comparison of specimens rather than markings. This is not the only topic in this text that contains some seldom-considered viewpoint or information, but it helps to set the tone. It is hoped that the content will appeal to practitioners, the legal community, and students who may be considering a career in forensics or another of the sciences.

1.1 Introduction

The fundamental building block of all organisms, deoxyribonucleic acid (DNA), is composed of a sequence of nucleotides unique to each individual. By obtaining the sequence (directly or by inference*) of a DNA sample and comparing it to a reference, the organism from which the DNA sample was obtained can be identified. The concept is familiar and is often used to identify

* Direct approaches entail obtaining and comparing gene (partial or full length) sequences or markers such as microsatellites. Indirect approaches include fingerprinting techniques such as restriction fragment length polymorphism (RFLP), amplified fragment length polymorphism (AFLP), or random amplified polymorphic DNA (RAPD) analysis.

a suspect using a blood sample. Given the frequent use of human DNA samples, and the popularity thereof in the media, it seems second nature to consider the forensic utility of human DNA. However, the utility of nonhuman DNA is burgeoning. For instance, canine DNA is in use in Israel to identify and fine owners who fail to clean up the feces from their dogs.[2]

Plant DNA is overlooked in many cases, in part due to a lack of recognizable boundaries concerning the usefulness of plant DNA (or forensic botany[3]); further, specimens are not routinely seized. The concept of obtaining samples from a crime scene, a suspect, or a suspect's belongings is dependent on knowing not only that the evidence exists, but also that, under certain circumstances, it can be rendered sufficiently useful for presentation in a court of law. For example, how often are plants or wood present at a crime scene? How often are fragments of wood present on weapons, vehicles, or suspects? What if those fragments of wood could be used to place a suspect at a crime scene? The answer to these, and other related questions, must eventually be answered to explore the utility of this concept.

Plant cells contain DNA in three organelles: nucleus, plastids (including chloroplasts), and mitochondria, all of which can be used in these studies. Plastid, genomic, and mitochondrial DNA have previously been extracted from commercial wood with a modified CTAB (hexadecyltrimethylammonium bromide) protocol (see the next section), allowing for species identification.[4] Plastid and nuclear DNA have been successfully extracted and amplified from submerged pine; however, this was successful due to modifications of the CTAB protocol.[5] Plastid DNA has also been isolated and analyzed from dry wood.[6] Likewise, nuclear, plastid, and mitochondrial DNA were isolated and amplified from dry oak wood.[7] Another group isolated and amplified plastid DNA from ancient (maximum 1,000 years old) wood samples.[8]

Thus, it is possible, using DNA analysis, to identify the source of a plant or wood fragment, and this could prove to be of considerable value in some cases. The work discussed here serves as a preliminary study testing the hypothesis that DNA sequence analysis can be used to identify the source of a plant or wood fragment.

1.2 Materials, Methods, and Results

Wood samples (0.01, 0.025, 0.05, 0.1, and 0.5 g) obtained from commercial cedar, pine, and poplar were used. The first step toward analysis of plant DNA is to isolate the DNA. To do this, the plant cells must be broken open; next, the DNA must be isolated from the remainder of the cellular contents and debris. A number of commercial kits and previously published protocols are available for DNA isolation. In this study, DNA isolation was done with a standard plant (*Arabidopsis*) protocol using CTAB.[9] Briefly, tissue was pulverized in liquid nitrogen and dried. A combination of chemicals and heat was used to extract nucleic acids (DNA and ribonucleic acid [RNA]). RNA was enzymatically degraded, and the DNA was specifically isolated, cleaned, precipitated, and finally dissolved in water. If successful, this DNA preparation contains the total cellular (genomic) DNA of the plant sample.

Once the genomic DNA has been isolated, the particular gene to be analyzed must be selectively amplified, which is done with a method called polymerase chain reaction (PCR). PCR specifically and exponentially amplifies the target DNA, allowing a sufficient amount of DNA for sequencing and analysis. To be useful for identification, the target gene to be sequenced must be a gene present in all plants and variable among individuals. This variability must be proportional to the degree of relatedness between individuals and species. For example, two individuals of the same species must have a more similar DNA sequence then individuals of two different species. In this case, a chloroplast gene (*trnS-trnfM*[10,11]) was used. Finally, the DNA sequence of the target gene would be compared to other sequences in databases or with other (reference) samples. From this, the identity of the organism from which the DNA was obtained might be ascertained.

Unfortunately, this study did not yield the anticipated results. DNA isolation, and consequently the amplification of the chloroplast gene, was not successful.

1.3 Discussion

While this preliminary study was not successful, others have previously obtained wood DNA, amplified plastid genes, and determined the identity of the organism (as discussed in the first section). A number of challenges are inherent in these studies; with troubleshooting, these may be overcome. Troubleshooting and protocol development may be specific to the plant species[6,12] and would need to be done by any forensic lab performing these analyses. As such, while we did not succeed, this analysis is possible.

Perhaps the most challenging step in extracting plant DNA is breaking open the plant cell walls. The use of commercial kits may help, but so will determining the appropriately small size of the starting material and the respective method of pulverization. For example, a major challenge in our work was the large size of the initial sample, which impeded pulverization in liquid nitrogen. In the absence of a sufficiently pulverized sample, an insufficient number of cells would be exposed to the chemicals. Consequently, an insufficient quantity of DNA would be obtained. It is also important to note that some DNA is inevitably lost as the DNA extraction protocol is carried out; therefore, a sufficiently pulverized or ground sample is an important starting point.

DNA loss occurs during any extraction and purification; however, this may be particularly detrimental when the sample is wood. This is partly due to the way in which trees grow. While bark is composed strictly of dead cells, the cambial layer (from which wood is formed) contains only live cells, sapwood is a mixture of living and dead cells, and the heartwood is composed completely of xylem or lignin (dead cells), containing no organelles (summarized in References 7 and 13). Given that the wood sample may contain a mixture of living and dead cells, a lower DNA yield is to be expected. In addition, DNA degradation begins once a tree is felled (as cited in Reference 6), which may prevent amplification of the gene of interest. Taken together, for maximum efficiency the gene for analysis should be smaller (or a partial gene or conserved marker) and present in a high copy number per cell.[7]

As mentioned, following extraction the DNA must be amplified for further analysis. A number of inhibitors may be present in the sample that decrease the efficacy of, or altogether prevent, amplification. These inhibitors may include tannins,[7] terpenes, and other impurities.[4,14] DNA oxidation may also prevent amplification.[4,15] While the effect of some inhibitors may be mitigated by the phenol component of the extraction protocol, the addition of chemicals such as polyvinylpyrrolidone (PVP[6]) or *N*-phenacylthiazolium bromide (PTB[4,15]) may also increase success. Sample contamination (e.g., by pollen or other foreign particles present in cracks within the sample) is also an important factor for consideration.[7]

A second major challenge is that DNA sequence databases and fingerprinting techniques may not be sufficiently developed (as cited in Reference 4) for plant species or even genus-specific identification. However, the question here is not to which species the samples represent but whether the reference sample and the evidentiary sample are the same. To answer this question, all that is needed are the two sequences and software to compare (align) the sequences.

While the utility of using plant DNA as evidence is theoretically possible and is relatively infrequently used, a number of issues and challenges will need to be overcome for the field of forensic botany to flourish. Careful attention must be paid to choosing the particular gene to analyze, especially with regard to the specificity of the chosen gene. For instance, can two or more individuals of the same species have the same sequence for that particular gene? If so, what are the ethical ramifications of using that gene to match an evidentiary and reference sample? Even if the sequence is unique to the individual organism (which is possible[7]), is it possible that, for example, two door frames or window sills in the same geographic area were made of wood from the same tree? As with any analysis, there are and will be limitations to this strategy. As such, more research is needed to fully investigate the advantages and limitations regarding the role of plant DNA in forensic science.

Acknowledgments

We would like to acknowledge the contributions of time, supplies and inspiration by many of those at Queen's University, in particular Dr. K. Ko and J. Powles (MSc). Thanks also to Dr. V. K. Walker and an NSERC (Natural Science and Engineering Research Council of Canada) grant awarded to her.

References

1. Botanical Society of America. BSA's Classroom. *Plant Science Bulletin*, Fall 2006, 52(3), 52–53. http://wwwbotany.org/PlantTalkingPoints/crime.php (accessed December 27, 2009).
2. Landau, A. 2008. An Israeli city is using DNA analysis of dog droppings to reward and punish pet owners. http://www.reuters.com/article/idUSLG37942520080916 (accessed September 16, 2008).
3. Coyle, H. M. 2005. *Forensic botany: Principles and applications to criminal casework*. CRC Press, Boca Raton, FL.
4. Assif, M. J., and C. H. Cannon. 2005. DNA extraction from processed wood: A case study for the identification of an endangered timber species (*Gonystylus bancanus*). *Plant Molecular Biology Reporter*, 23, 185.
5. Reynolds, M. M., and C. G. Williams. 2004. Extracting DNA from submerged pine wood. *Genome*, 47, 994.
6. Rachmayanti, Y., L. Leinemann, O. Gailing, and R. Finkeldey. 2006. Extraction, amplification and characterization of wood DNA from Dipterocarpaceae. *Plant Molecular Biology Reporter*, 24, 45.
7. Deguilloux, M.-F., M.-H. Pemonge, and R. J. Petit. 2002. Novel perspectives in wood certification and forensics: Dry wood as a source of DNA. *Proceedings of the Royal Society of London B*, 269, 1039.
8. Leipelt, S., C. Sperisen, M.-F. Deguilloux, et al. 2006. Authenticated DNA from ancient wood remains. *Annals of Botany*, 98, 1107.
9. Murray, M. G., and W. F. Thompson, 1980. Rapid isolation of high molecular weight plant DNA. *Nucleic Acids Research*, 8, 4321.
10. Shaw, J., E. B. Lickey, J. T. Beck, et al. 2005. The tortoise and the hare II: Relative utility of 21 noncoding chloroplast DNA sequences for phylogenetic analysis. *American Journal of Botany*, 92, 142.
11. Demesure, B., N. Sodzi, and R. J. Petit. 1995. A set of universal primers for amplification of polymorphic non-coding regions of mitochondrial and chloroplast DNA in plants. *Molecular Ecology*, 4, 129.
12. Weising, K., and R. C. Gardner, 1999. A set of conserved PCR primers for the analysis of simple sequence repeat polymorphisms in chloroplast genomes of dicotyledonous angiosperms. *Genome*, 42, 9.

13. Kramer, E. M. 2006. Wood grain pattern formation: A brief review. *Journal of Plant Growth Regulation*, 25, 290.
14. Shepherd, M., M. Cross, R. L. Stokoe, L. J. Scott, and M. E. Jones. 2002. High-throughput DNA extraction from forest trees. *Plant Molecular Biology Reporter*, 20, 425a.
15. Poinar, H. N., M. Hofreiter, W. G. Spaulding, et al. 1998. Molecular coproscopy: Dung and diet of the extinct ground sloth, *Nothrotheriops shastensis*. *Science*, 281, 402.

Signs of Evolution 2

2.1 A New Beginning for Forensics

2.1.1 Forensic Evolution

The debate about deficiencies of forensic methods with respect to their accuracy, credibility, and integration into the criminal justice system has intensified.[1] Installation of an independent national institute for forensic science in the United States would address the need for validation of existing technologies, protocols, and procedures to establish forensic science standards.[2] These standards can be expected to promote the rapid evolvement of forensic science into a mature discipline in the many countries currently evaluating their methods.

Today's best practice is tomorrow's historical anecdote. The need to seek the best methods and equipment is obvious, particularly when the evidence may vary in stature from meters to nanometers and observations are made in diverse natural settings. This does not mean that simple techniques will not continue to play a role. Apart from their diversity, the principles related to forensic science are often simple once understood; this chapter seeks to provide a starting point.

Science involves systematic experimentation and observation with the goal of explaining natural phenomena. It is crucial that the tools used to observe the outcome of experiments must possess a high degree of reliability. Reliability and accuracy can be obtained through calibration of the equipment used, such as interlaboratory calibration. This requires adhering to generally accepted and consistent theories and standards, which should take the form of validation procedures that must extend beyond the confines of the lab.

Each measurement taken from evidence contains a degree of uncertainty due to the limitation of observations, tools, sampling techniques, and systematic problems introduced by the people using them. Conclusions, based on experimental results, need to include an understanding of the vulnerabilities involved. Reference to measurement provides credibility for estimations concerning both precision and accuracy, which can be further supported by statistical analysis, thus improving testimony about the nature of observations (see Chapter 5).

The opportunity to see exhibits in their context provides substantial information regarding the activities and processes that have affected the scene. When the context is missing, errors or bias can occur, particularly when scientists are asked to evaluate evidence removed from a scene. This is referred to as the "investigator/evaluator dichotomy."[3] To prevent this bias, the judgment of a forensic scientist needs to be embedded in a "structured and disciplined environment (i.e., a large scale program of collaborative studies and proficiency tests)."[4]

2.1.2 Effects of Scientific Evolution

There are no such things as applied sciences, only applications of science.[5]

The word *science* translates with no distinction or boundary regarding particular fields of study, but merely refers to the principled investigation of all manner of interests. It is reasonable, given this holistic definition, that "the scientific method" is merely a reference to following generally accepted values and practices. The scientific method is the pursuit of reliable observations and justifiable proofs rather than a single methodology that is set in stone. Each crime scene, every object in it or related to it, can be treated as a scientific experiment in which past observations, techniques, and methods will be tested against this new set of challenges.

Interdisciplinary work often blurs the distinctions between "traditional" sciences. Each division of a science benefits from some common ground with other disciplines. The growing depth and complexity of the respective disciplines of scientific knowledge makes it impossible to become an expert in all of these pertinent disciplines simultaneously.

The future evolution of forensic science will see rapid development of its research base, relying on scientific principles and methodologies across the aforementioned multitude of disciplines. The ability of an investigator to locate and treat evidence is currently developed by experience and enhanced by understanding that science governs the physical and mechanical properties of the substances involved.

Science provides clarity to many topics, thus building a template for further study. Groundwork has been laid by research and experimentation in many areas, such as the role of liquids other than blood (see Chapter 6). Scientific research conducted until now has not adequately addressed all forms of impression formation or deformation, and there is a great value to be found in the practice of re-creating the phenomena in question using simple experiments.

Exaggerated versions of events are useful to illustrate a concept, but not all variations of similar origin will be as easy to detect or be confined to the same boundaries. Empirical testing is undertaken to avoid rash assumptions. The behavior of materials in reaction to stress and the resulting deformation can be complicated. These matters relate directly to the study of contact mechanics.

Legal communities have an obligation to integrate the fabric of changing disciplines and the evolution of forensic evidence. The legal system wanted (actually demanded) this change, and now that the result is taking effect, the legal community needs to keep pace. The use of research must be appreciated, and new methods will need to be embraced, whether supported by statistical calculations or simple logic.

The first step toward understanding begins with the simple observation that impression evidence is, by nature, fragile. Forensic disciplines have strived to improve techniques and practices, widening the scope of their abilities with each new finding. The demonstrated vulnerability of markings offers correspondingly important properties that can assist in determining the viability of a marking to fit within specific events. Drying water marks and abrasion of a substrate are typical anomalies for which the courts need to develop a better understanding.

Workloads can be forecast to increase at all levels of involvement. Management of the workforce and caseloads and minimization of bias will be growing tasks. Perhaps the greatest successes will be rooted in more comprehensive use of each forensic discipline wherein the values, observations, and findings can be held in greater esteem by the courts.

2.2 Substrate to Structure

The most basic of introductions to all of the evolving sciences make the point that all matter consists of atoms. Basic concepts further state that atoms are composed of an array of neutrons, protons, and electrons. This fundamental arrangement is the blueprint for matter of all kinds, from living organisms to the rocks, earth, and liquids that they inhabit.

All forces that affect matter are in turn acting on, and reacting with, atoms. Atoms are the key to studying the organisms of biology, the physical forces of engineering, and the chemical and electrical properties of chemistry and biochemistry. The forces and conditions that bind and influence the behavior of atoms create elements of more complex structures or situations.

The physical states of matter help to define each substance. Even the most stable of solids can be melted or decomposed provided the temperature is high enough. Also to be considered is the physical or chemical effect of one substance on another.

To appreciate how materials interact and behave, it is useful to learn about how they are held together. Covalent bonds, for instance, are known to bind the "like" atoms of diamonds or graphite as well as the "unlike" atoms of carborundum and quartz.

Natural systems and structures are often populated by recurring themes and varying levels of complexity. Conservation of mass, conservation of energy, symbiotic relationships, reciprocal conditions, and symmetry are but a few examples. Forensic disciplines are similarly populated by many recurring themes.

Forensic literature generally treats substrate as though each material exists in a constant but unquantifiable state. In fact, a substrate is part of a greater environment where other substances and a number of forces or attractions intersect at a given location in time. The exact nature of any substrate is dependent on its surrounding influences and can neither be taken as a constant simply because it displays similar features and appearances nor have its properties ignored.

Flooring or decking, for instance, will have a predictably lower coefficient of friction when wet than dry; this is obvious. If there is a significant difference in relative humidity, the same flooring may appear to be unaffected, but the difference in the coefficient of friction is measurable. One must be able to take into account the effects of changes in the environment, whether the scene is indoors or outdoors, wet or dry, worn or new, as each set of variables can have dramatic effects.

When wet, grass can behave like a slide, making it impossible to maintain one's balance. An exterior porch, wet with dew, creates another slip-and-fall hazard. Yet, one must keep in mind the possibility that what appears to be an innocent event may have been accelerated by a push or shove initiated by someone intending to take advantage of the naturally occurring hazards.

When substrates combine a variety of complexities such as exchanges between fluid, solid, and gas interfaces, they may be more appropriately studied as structures. The initial substrate will contribute a number of base characteristics, which are further complicated or modified by internal kinetic energy (energy of motion) or conditions such as adsorption. The point of concern is that a study of a substrate is incomplete without consideration of the greater structure.

Characteristic forms of structural integrity may include expansion, contraction, bending, stretching, dimensional stability, or shear, as found within the particular structure of interfacing substances (with the added expectation that there may be a cyclic effect).

Consider that few impressions are found that precisely imitate a known impression. Crime scene markings are routinely altered by a variety of conditions that separately or in combination often defy interpretation. Effects such as the almost-constant presence of adsorbed liquids on common surfaces may become defined, explainable, or easier to interpret as a result of empirical studies.

2.3 Forces

2.3.1 Forces Big and Small

We expect to see a significant difference in the appearance of an impression under a greater load. The fact that we expect things to happen in a predictable fashion is not unreasonable; it just is not always applicable.

The familiar question of whether you would rather have the foot of an adult elephant or the high heel of a 120-lb woman step on you is easily calculated (with some basic data about the elephant, the foot, and the size of the high heel, one can easily see that the weight of the elephant is comparatively more evenly distributed). This example serves to illustrate the matter of comparisons between conforming (exact fit, as in flat to flat or similarly curved) and nonconforming surfaces (with at least one surface somewhat irregular) or variations of the two.

Most forensic impressions are the product of contact between largely nonconforming surfaces. Friction ridges, the lugs of patterns, or features of a tool are routinely found to vary in the degree of conformity between the contacting surfaces according to load. Engineers are not as concerned with the nature of resulting impressions as they are intrigued by the interplay of effects and commonalities that subsequently allow the forces to be generalized (as in the determination that the movement of meshing gears involves quantities of both rolling and sliding forces).

Take, for instance, a doubled amount of pressure with any type of impression. On a relatively soft substrate, our expectations will be met with a fairly proportionate result. A harder substrate will also yield a proportionate result, but it is not visibly proportionate in that there may be only a minor or possibly imperceptible deformation. Our perceptions and expectations can lead us to expect impressive results or even (in the case of minute changes) to dismiss that which did occur.

When ability of a material to retain its shape is exceeded, the amount of force required to sustain a continuance of change can be significantly less than was required to overcome the initial resistance to that force. An example of this was encountered with the controlled bending of threaded 1/4-in. carriage bolts during research for this book (Figure 2.1). The bolts, mounted in a vise, were exposed to pressure applied at approximately 45°.

The amount of force was measured by transferring the pressure through the bolt to the scale. For each bolt tested, a small amount of deformation was observed in the range of 25–30 kg, and failure was observed between 60 and 65 kg. Once the material (bolt) had begun to fail in each case, significantly less pressure (about 10–20 kg less) was required to sustain the rate of deformation.

Forces and the conditions they impose during the creation of impressions can be expected to be present in each phase of contacts, and most

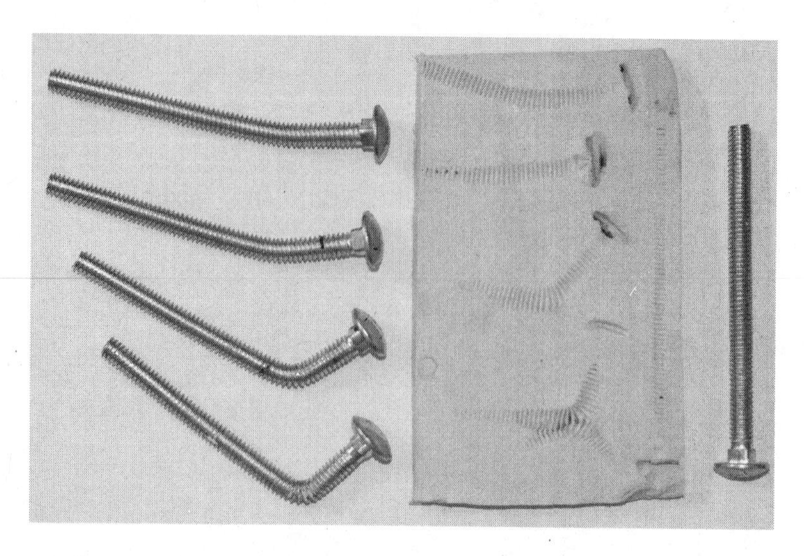

Figure 2.1 Bolts affected by stress. In this image, the bolts were systematically bent, and the difference in the thread pattern near the bend was recorded in a cake of modeling clay.

impressions involve more than a single force. These forces are present because they typically form and shape the world around us. The visible forces are well known and, for the most part, easily recognizable. It is frequently the smaller, lesser understood forces and conditions that are focused on here.

Thinking about forces requires some consideration of simple relativity. A bullet that reaches a target will often penetrate that target. The same target traveling at almost the same speed as the projectile is far less likely to suffer damage (in keeping with Newtonian motion).

Every child soon learns that gravity is the force that keeps our entire planet in its orbit and prevents us from flying off the surface and into space. There can be no doubt that gravity is a significant force. It is less well understood that the effects of gravity are not comparatively strong and vary greatly in effect according to the situation.

The effects of gravity are obviously proportionate and must be considered in a different manner at the molecular level. Gravity is mass dependent, and other forces that are far less visible to humans can overtake the effects of gravity as the mass of a subject decreases[6]:

> You can drop a mouse down a thousand yard mine shaft; and, on arriving at the bottom, it gets a slight shock and walks away. A rat is killed, a man is broken, a horse splashes. For the resistance presented to movement by the air is proportional to the surface of the moving object. (p. 55)

2.3.2 Plainly Small Forces

Small forces often appear relatively weak, but they can have significant effects. Adsorption (the binding of molecules to a surface, as opposed to molecules that are absorbed into or toward an interior location) is a small thing that (in part) can account for the ability of a mighty tree to move water from its roots to its leaves. Adsorption also plays an important role in the condition of surfaces and is only temporarily absent inside a vacuum, an effect that can be observed if one tries to examine affected exhibits by vacuum metal deposition techniques.

Miniscule forces also play a role in the ability of one surface to grip another. Deformations can be compounded by the effects of small intermolecular forces. One can be sure that examples of the involvement of these forces could be found to influence the components of each particular instance when an impression results.

The strengths of intermolecular forces vary greatly but are much weaker than other bonds. This is demonstrated by the fact that it generally takes more energy to break ionic or covalent bonds than to effect a change of state. A lower or higher boiling point (or melting point) of one substance compared to another substance will be an indication of correspondingly weaker or stronger intermolecular forces.

There are three main types of attractive intermolecular forces (secondary bonds) between neutral molecules: dipole-to-dipole force, London dispersion forces, and hydrogen-bonding forces, which are collectively known as van der Waal's forces. Another, ion-dipole force, applies to solutions. "All four forces are electrostatic in nature, involving attractions between positive and negative species. All tend to be less than 15% as strong as covalent or ionic bonds."[7(p446)]

Ion-dipole forces can be represented by the solution of the ionic substance (a primary bond) NaCl in water. This description of salt water depends on understanding the attractive forces between both the positive and negative ends of the ion to the oppositely charged ends of the polar molecules (dipoles).

Dipole-dipole forces are then the attraction to oppositely charged ends of two neutral polar molecules. Although weaker than ion-dipole bonds, these forces come into play when the molecules are close together. A clearer description of what happens with dipole forces comes from consideration of the movement of the charges.

Remember that electrons are often thought of as existing in a cloud. As the electrons are attracted at a moment in time by the presence of an opposite charge, an increasing number of them will gravitate toward the source of attraction. This movement of electrons creates a dipole moment in which the molecular forces are stronger in one area than another.

The importance of such fine distinctions becomes apparent with the polymers in outsoles, for instance. Weaker bonds at a given dipole moment can account for the introduction of scission, resulting in shrinkage or swelling. This process relates directly to the accumulative escalation of vulnerability for such materials with continued exposures.

2.4 A Basic Look at a Popular Substrate

2.4.1 Polymers

The family of materials most often represented as substrate found at crime scenes must be polymers. The subdivision into groups begins with natural and synthetic polymers. Natural polymers include wool, cellulose, leather, silk, and natural rubber. These substances are common to a wide range of elements in each crime scene.

The word *polymer* (many parts) is a composite that stems from the Greek language. The name reflects the structure in which many *monomers* (single parts) are combined on a molecular level to form a final product. The synthetic process of creating polymers (polymerization) is familiar to almost all forensic practitioners who create polymers by cyanoacrylate fuming techniques. Even the use of catalysts and inhibitors to control polymerization are at times employed in the development or enhancement of forensic evidence, such as the use of sodium hydroxide to accelerate and distribute cyanoacrylate fumes.

These analogous examples lead to some understanding of the way in which polymers are formed. Understanding the current use and potential future uses of polymers is another matter. Certainly, the construction of modern footwear and tires is representative of current polymer usage, but there are less-obvious examples that include medical implants, contact lenses, and electrical semiconductors.

Increases of waste in landfill spell out the need to reduce and eliminate packaging and the generation of manufactured goods. The need to reduce our use of oil-based substances is accompanied by advances in the fields of "green," renewable plastics (largely made from plant products rather than petrochemicals), biomedical polymers, and even nanotechnological applications.

Given the many reasons to expect a continued presence of polymers, it is likely wise to understand how species of the sort are made, function, and are degraded.

2.4.2 Primary and Secondary Bonds

Many clear polymers exposed to sunlight will turn cloudy or yellow over a relatively short period of time with no application of physical stress or force other than photodegradation. These products are obviously not designed to resist the energy delivered by photons, especially in the ultraviolet (UV) range of the spectrum. These rays are strong enough to break the bonds of the larger polymer molecules into smaller molecules, thus reducing strength and further escalating change to other properties of the substance.

Bonding is a cohesive arrangement of molecules within a particular substance[8]:

> The five types of bonds holding matter together may, broadly speaking, be classified in two main groups. The stronger ones, often denoted as primary bonds, have energy content of the order of 100–200 kcal/mol; for the weaker ones, the secondary bonds, the energy content is of the order of 0.1–10kcal/mol, or somewhat higher. (p. 14)

Primary bonds include ionic bonds such as sodium chloride, the covalent bonds of diamonds, and metallic bonds as found with sodium molecules. The sharing of electrons in primary bonds is a strongly cohesive arrangement. Metals share all the outer electrons, giving them a freedom of movement that accounts for their conductivity while yet ensuring the cohesion of the molecules.

Secondary bonds are found in the weaker substances, such as many polymers. While any bond can be subject to degrading effects, it generally takes less energy to rupture secondary bonds, giving rise to a variety of deformations.

Polymer design addresses the needs of particular industries by combining materials and additives that make useful contributions toward controlling problems such as the effects of UV light on polymer bonds. Carbon is an additive that absorbs much of the heat energy from light rays before they can affect molecular bonds. Carbon, in the form of a ground particulate (powder), is added to products such as vehicle tires, which are both strengthened and protected by the combination.

2.4.3 Manufacture and Growth of Polymers

We have introduced various types of bonds and forces. The following examines how polymers use those forces to form or are affected by them.

Addition and condensation are the primary methods by which most synthetic polymers are created from separate monomers. The reaction (with condensation) involves the removal of an unlinked monomer such as a water molecule from a monomer, which subsequently bonds to another short monomer and thereby increases the length of the polymer chain. This is a dehydration synthesis (the opposite of hydrolysis) catalyzed by a polymerase enzyme with natural polymers and effected by applications of acids and bases in the manufacture of synthetic polymers.

While it is not important within this text to offer an exhaustive guide regarding how commercial polymers are synthesized, it is useful to know some characteristics of polymers that most directly affect the appearance of impressions. A useful division of the many substances is provided by the manufacturing method (the materials that are present in recycling schemes are dominated by addition polymers), which in turn can be recognized by recycling symbols printed on the material. The most prolific symbols contain the abbreviation of the material name, a numeric code, or both.

The most common American recycling codes are as follows:

1. Polyethylene terephthalate (PET or PETE), common in packaging and some bottles
2. High-density polyethylene (HDPE), common to plastic films and shopping bags
3. Polyvinyl chloride (PVC), generally a clear and pliable film that is also used in pipe fittings
4. Low-density polyethylene (LDPE), also used in blown films and packaging
5. Polypropylene (PP), from which kitchenware, fibers, and some appliances are made
6. Polystyrene (PS), used as insulation and disposable food containers

All of these are examples of addition polymers. The easiest of these examples to visualize is the "opening" of the double bonds of ethylene molecules to connect to other ethylene molecules, as shown in the model of a single ethylene molecule.

Model of a Single Ethylene Molecule

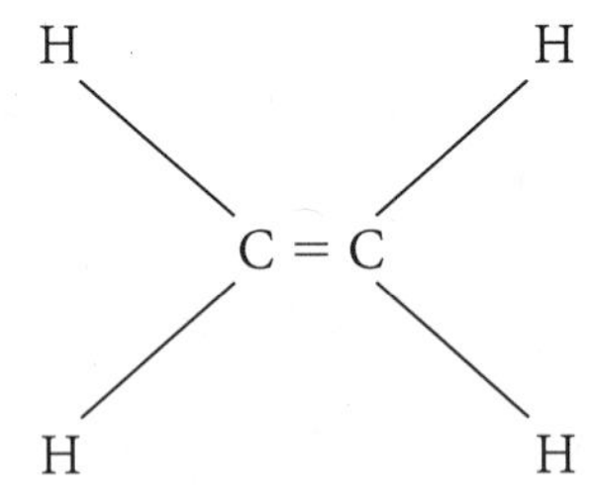

Note the double horizontal lines between the two carbon atoms. Those lines represent a doubled bond between the carbon atoms. To create the polymer, polyethylene, those double bonds are opened to create shared single bonds between a series of carbon atoms (still bound to two hydrogen atoms), thus resembling a chain of railway cars (see the model of a polyethylene chain).

Model of a Polyethylene Chain

```
    H   H   H   H   H   H
    |   |   |   |   |   |
  — C — C — C — C — C — C —
    |   |   |   |   |   |
    H   H   H   H   H   H
```

Other polymers have more complex structures. Polyethylene has a high molecular mass and can be reshaped with applications of heat and pressure (thermoplastic properties) into other products. Polymers composed of more than one type of monomer are known as copolymers. Thermosetting polymers, by comparison, form solid but permanent objects such as "Bakelite" materials, which are not reversible and cannot be recycled.

2.4.4 Natural Polymers and New Products

The use of natural polymers changed dramatically in the last century with applications of vulcanization and the discovery of synthetic polymers. Issues, including health risks and ecological concerns, are causing researchers to review the efficiencies illustrated by natural substances.

Research has yet to identify precisely how natural polymers such as cellulose are generated by plants. One of the most recent studies regarding the creation of plant cellulose suggested that some previously not well understood organelles, commonly found in plant cells, act as a general scaffold for organizing the cell walls. These organelles, microtubules, play a role in the growth of plant cells from the inside.[9]

The process of natural polymer growth is of great interest to researchers, who are attempting to perfect substitutes for the many polymers currently produced from hydrocarbons. The green aspects of natural polymers and rising cost of oil spell out the need for change. In forensics, this means that by becoming familiar with the new materials, it will be possible to keep pace with some knowledge of the characteristics that affect impression evidence.

New research is looking in much closer detail at cellulose[10]:

> Cellulose has a complex, multi-level super-molecular architecture. This natural polymer is built from super-fine fibrils having diameters in the nano-scale, and each nano-fibril contains ordered nano-crystallites and low-ordered nano-domains. (p. 1403)

The world of forensic detail is still largely limited in size to macro photography, including a scale divided into millimeters, but as technology advances, the impact of features measured in micrometers and even nanometers is becoming significantly more important. Semiconducting nanotubes were first separated from metallic carbon nanotubes in 2003. "Carbon nanotubes have also been spun into fibers with polymers, adding great strength and toughness to the composite material."[7(p518)]

This integration of new technology will have the capability of introducing a range of materials of far greater smoothness than glass. These changes to substrate will result in a visible increase in clarity with some impressions. Applications that affect biometric endeavors may offer potential improvements in the way a friction ride marking is recorded. Perhaps of more concern is the greater slipperiness, which will likely call for engineered texture to control the new coefficient of friction for such materials.

Industrial control of frictional forces will be one of the first benefactors of current research. One can immediately appreciate the usefulness to marine, aeronautic, and automotive designers in improving the efficiency of their products in motion, the ability of roofing materials to survive hurricanes, or net gains in power derived from wind turbines. Study of natural gripping systems as demonstrated by crickets, geckos, and snake skin will offer inspiration and even improvements on natural systems, some of which are already appearing: "Laser texturing has been employed to create microgrooves and microdimples that act as lubricant reservoirs."[11(p1340)]

2.4.5 Polymer Strength and Deformation

The creation and effects of deformations are of particular significance in many of the subsequent topics in this text. Expert witnesses must provide interpretations that explain the presence of distortion with as much detail and clarity as any other form of evidence. This emerging line of study includes reference to linking the internal and interactive properties of materials under stress with a grasp of the dynamics and kinetics that define the stress.

The field of material deformation, also termed *rheology*, is the study of a changing relationship between a subject material and its environment. Minor changes, which may or may not be readily noticeable, can be a factor in many classes of deformation. The amount and precise effects of deformations can be calculated by material designers, who select from a range of substances that are suited to a set of specifications.

Designers have used a number of options to change the physical properties of specific polymers. PVC, used to make fairly sturdy pipes for the plumbing trade, can be modified using plasticizers (with a lower molecular mass) to make the material sufficiently pliable for the production of rain boots.

Natural rubber is a familiar material that, without the addition of sulfur and heat (vulcanization), is far too soft to be of much commercial use. In the vulcanized form, a wide variety of products are made from rubber, and further additives can extend that range. Vehicle tires, for example, are made stronger and more resistant to UV degradation using additives. If the composition of substances, solvents, or characteristics is unknown, any interpretations are usually limited to observations. Comparison of a crime scene marking must therefore entail an amount of research-based knowledge that allows an examiner to make some useful conclusions.

Physical forces may include torsion, tension, or compression, and the responses they generate may include fracturing, bending, stretching, or shear. Once inertia has been overcome or a substance reaches the shear point, flow or motion results. The aforementioned are commonly associated with the properties of an outsole or a tire, but the same types of forces also pertain to the soil in which the marking is found.

The ability of a substance to react to a force in a particular way depends both on the force and on the types of bonds that make up the substance. Each example of the particular behavior of a substance is a class characteristic known as a property, which can be used (in part) to define the substance. Temporary and permanent deformations and other measurable reactions to force are the substance of rheological studies, such as of the degradation of polymers.

2.4.6 Hysteresis

Hysteresis falsely implies a topic of great complexity. The term explains effects imposed on an impression not only by forces, but also by the nature of combined surface pairings. This single topic could offer enough substance for another entire book.

Hysteresis is defined as a slowed return to an original shape or form. This is a condition observed in other disciplines, such as electronics, and is best illustrated as a forensic matter by study of elastomers. Hysteresis in this sense applies to the tendency of a rubber band to return slowly to its shape on the release of a stretching force. Friction skin and metals are not made of rubber, but both will demonstrate examples of hysteresis under pressure.

The response of a shoe or tire to the small asperities on a fairly smooth substrate is similar to the response illustrated by a rubber band. This phenomenon, termed *microhysteresis*, is on a much smaller scale. The interface of a softer rubber or polymer with a rougher asperity can result in a portion of the rubber tending to conform to the shape of the asperity. This creates a condition that allows the incidence of grip, which can explain a vital part of the mechanism of friction[8]:

> During the elongation of vulcanized rubber, a process which is reversible to a high degree, the segments of the chain molecules slide along each other. However, this flow is limited by the linkages which exist between these molecules at certain points, so the system is both viscous and elastic. As a consequence of the interaction of adjacent molecules during their viscous flow on

> stretching, frictional heat will be developed. This is the irreversible part of the energy of stretching that is the hysteresis. Keeping the rubber stretched for a longer period of time results in additional viscous flow, manifested by a relaxation of tension. (p. 223)

Studies of hysteresis and the limits of elasticity, as applied to the behavior of polymers, are a simple matter that is part of greater topics. Mechanisms of grip, friction, wear, and material failure begin with models of hysteresis on a much smaller scale. Microhysteresis becomes much easier to comprehend with modeling based on the behavior of visible materials such as rubber bands or even balloons.

In a simple experiment with three rubber bands immersed in diesel fuel for a period of 50 minutes, the bulk material swelled respectively between 12% and 17% as measured both under load and at rest. Control samples exposed to sunlight for 8 hours showed no measurable change. It is reasonable to consider that further testing could provide some data that may reflect a weakening of the materials involved, as discussed in Chapter 10 in relation to experiments performed on footwear outsoles.

2.4.7 Elastic Limit and Deformation

The distension of a material past the elastic limit for that material results in deformation. The more brittle materials, such as metals or glass, will develop fractures or even break, but polymers and rubbers tend to behave a little differently before shear occurs. In a drastic example, rubber and similar substances tend to change size, either locally as in the bulges seen in some tire sidewalls or more widespread as with the overextension of a rubber band or a balloon that will no longer return to their original shape. This will then take the form of a permanent hysteretic deformation.

Solids as firm as rock possess some degree of elasticity. The common image of a substance failing to obey Hooke's law is most often represented by the upper limit at which a material breaks, fractures, or shears. Between the extremes of hysteresis and fracture is the realm of hysteretic deformation.

Careful measurement of a rubber band that has been stretched past its elastic limit results in measurable permanent deformation. This same effect was observed in separate tests conducted with both rubber bands and rubber balloons.

Six white, round balloons, all of the same make and brand, were divided into two groups (Figure 2.2). The first three balloons, in a restive state, were marked with 12 dots arranged in 1-mm intervals. The next three balloons were inflated to an approximate maximum diameter of 30 cm. A set of 12 dots was drawn on these balloons at 4-cm intervals.

The first three balloons were inflated to an approximate maximum diameter of 30 cm. These balloons were subsequently deflated and measured. The

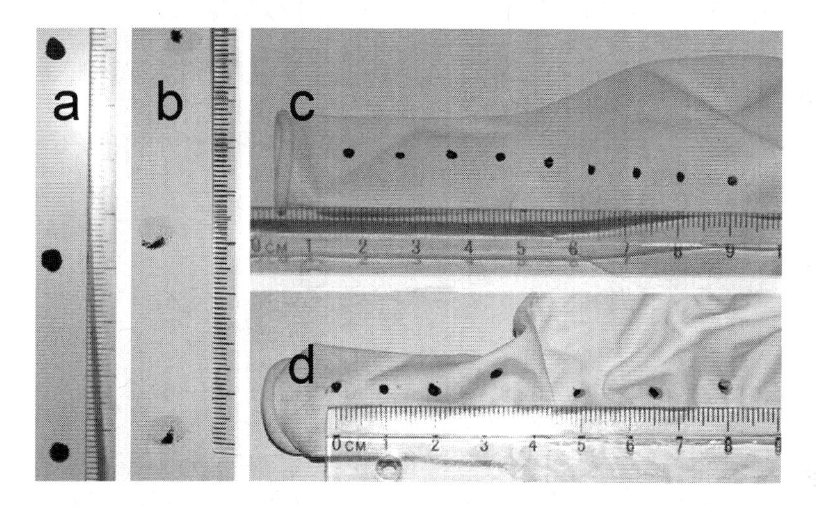

Figure 2.2 Deformation of balloons. These four images illustrate that there can be a proportionate effect on the material that relates directly to the amount of load applied or relieved: (a) a set of dots drawn on an enlarged specimen at 4-cm intervals; (b) the product of inflating the specimen in Figure 2.2c; (c) a set of dots drawn on a different balloon at 1-cm intervals; (d) the product of the deflation of the specimen in Figure 2.2a. A more detailed experiment (designed but not implemented for this text) included the effects of time and variations in the amount of pressure on the results.

second set of balloons was deflated and measured. The graph in Figure 2.3 shows the average of measurements taken for the two conditions.

While the number of scenes that will require an examination of balloons is likely to be a small percentage, the illustration serves to depict a characteristic of elasticity that may need to be considered in other cases that include vestiges of elastic behavior.

2.4.8 Furthering an Understanding of Polymer Deformation

Addition polymers, as mentioned, may be reshaped. The application of heat and pressure to these substances does not damage the material. Degradation of either addition or condensation polymers can result from moments of vulnerability to UV light, exposure to a solvent, and in response to an applied mechanical force.

The reduction of at least some polymer chains into the constituent monomers can occur during a process known as *hydrolysis*. This is a fairly simple reversal of the process of condensation (by the process of dehydration synthesis) by which a water molecule is introduced to the covalent bond of the polymer, thus breaking the polymer chain into monomers. This process is also facilitated by an enzyme with natural polymers and acids or bases with synthetics.

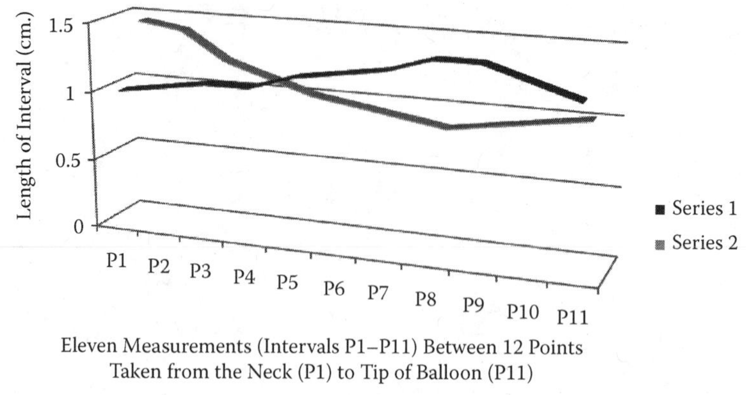

Figure 2.3 Change imposed by the effects of inflation and deflation. The two series of results tend to correspond to those areas of the balloon that were most and least changed by inflation and deflation. Intervals toward the neck of the balloon (left) and the tip of the balloon (right) appeared to respectively exhibit less change than the areas corresponding to visibly greater expansion.

Water is not the only solvent that can cause weakening between the intermolecular bonds of a polymer chain. The well-known rule for estimating the potential for this scission to occur is that "like dissolves like." The nature of the molecules, and the solvents that affect them, is one of the fundamental differences between polymers of biological origin versus those that are manufactured.

Polymers entering a mold may create incidental markings, as in the rippled effect shown in Figure 2.4. New outsoles have been observed to exhibit what appear to be Schallamach patterns (wavelike surface characteristics attributable to wear) that are more likely a product of the temperature and condition of the mold at the time the outsole was formed or the presence of additives such as plasticizers or mold release agents. Similarly, interesting patterns are often found with manufactured items such as the molded polyethylene container, apparently covered in enormous friction ridges, pictured in Figure 2.4.

2.4.9 Solubility

It is common knowledge that many polymers, with varying degrees of resistance, are soluble; in fact, the classification of a substance as insoluble does not rule out some small degree of solubility. Dissolution rates depend on both physical properties of the given substance and the nature of their chemical structure, such as polarity, molecular weight, branching, degree of

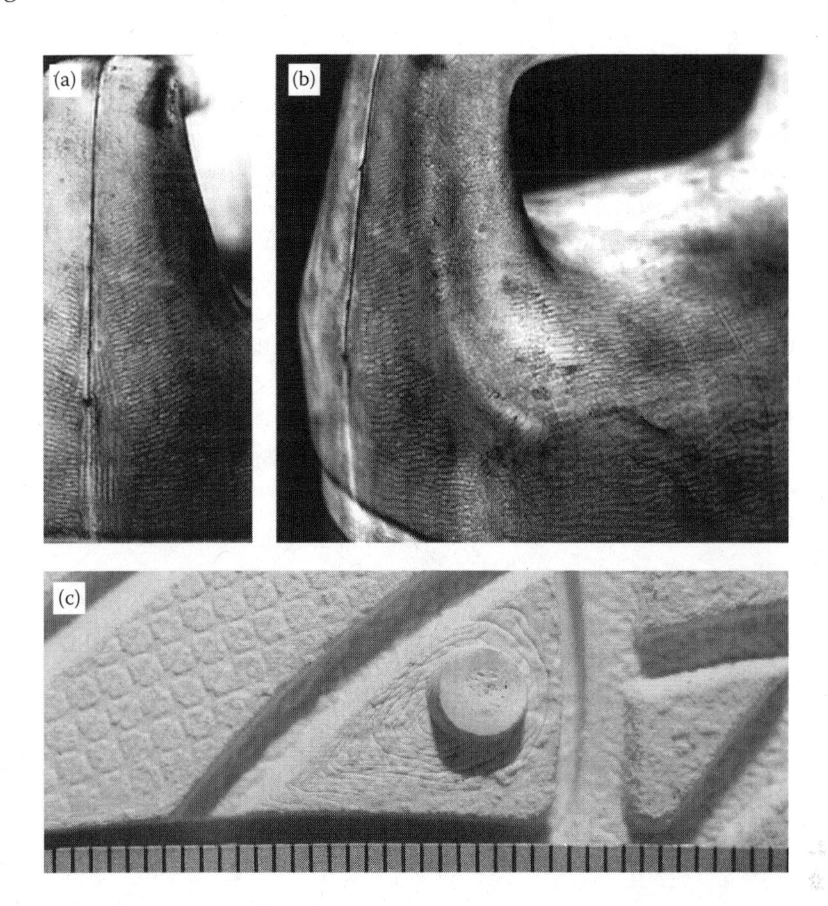

Figure 2.4 Synthetic markings. Two views of vary large ridge patterns are shown on the same automotive washing liquid jug in (a) and (b). The size of the ridges suggests that they may have been a product of handling prior to manufacture. This theory is discounted by an absence of features typical of a palm impression such as creases, variance in ridge flow direction, or the presence of deltas. The bottom portion of the figure (c) illustrates the effect of mold temperature on flow characteristics (seen as assorted wrinkle-like features around an injection point).

cross-linking, and crystallinity. The general principle that like dissolves like is both evident and prevalent with many polymers.

Nylon is a crystalline polymer that is more polar than other polymers (e.g., polyethylene). Nylon can be dissolved at room temperature by solvents with the ability to react with its chains. Hydrogen bonding is an example of how a polar solvent can dissolve a polar solid (over time).

When a low molecular weight substance such as sucrose is exposed to water, the rate of dispersion into a solute is almost immediate. A synthetic polymer is designed to be more resistant to solvents, and that dissolution process can take days or weeks of exposure to produce noticeable results. In an experiment conducted with a nonpolar footwear outsole, the exposure to three solvents over a 10-day period (continuous exposure of 240 hours) resulted in a series of widely differing results at or near room temperature.

In each test case, the subject material was observed to swell more in one direction than another. This observation is in keeping with the nature of polymers, which will react differently depending on the method of manufacturing. Some polymers are intentionally stretched or cross-linked to increase the ability of the polymer to resist the intrusion of solvents.

Solubility characteristics can be expected (with an appropriate combination of solvent and solid) rather than predicted with any confidence. Observations of a general nature allow some insight regarding the solubility of a substance. There are no rules but only some general guidelines: "For example, experiments show that all common ionic compounds that contain the nitrate anion, NO_3^-, are soluble in water."[7(p127)]

2.5 Symmetry

2.5.1 Symmetry by Design or Circumstance

Polymers and the products they are used to make can express a great deal of symmetry. With many manufactured products, symmetry is generally a desirable and therefore a common feature. Manufacturing processes contribute to a symmetrical outcome, and some examples are produced according to narrow tolerances.

Symmetry found in impressions is expected in footwear, tool marks, and the highly sophisticated designs of tires. Symmetry in what could otherwise be a random marking in impressions can be a warning sign that the marking may be afflicted in some way.

A *double tap* refers to a situation often found with exhibits populated by numerous imprints, such as a beverage container. When subsequent impressions overlap each other, there is a good chance that portions of two friction ridge imprints can appear to merge. Areas of high traffic or impression density (for any type of impression) can be expected to be inhabited by varying degrees of overlapped impressions, which will require careful interpretation.

While double taps refer to any overlapping of two marks, the merging aspect is a little less common. Research for the article "The Significance of Butterflies"[12] involved the creation of thousands of impressions, one beside the other, in attempts to create representative examples of fairly high quality. The resultant images resembled a collection of butterflies on a page (hence the title).

In creating the illustrations, it was found that about one in a dozen attempts (on average) would result in an area of overlapped friction ridges that merged in a fairly convincing manner. It was viewed that those areas could prove misleading if other clues (such as the apparent butterfly shape) were not present. Friction ridges were successfully caused to merge between different digits and different individuals as well as friction ridge imprints between plantar and palmar regions for which the ridge widths corresponded.

The study of merged impressions, no matter what name is used, demonstrates the potential for a commonly handled object to possess such features as a matter of chance. A practitioner will logically respond to an area that has been subjected to multiple handling by searching for identifiable markings of sufficient area to permit the formation of a conclusion. The examination needs to encompass those areas adjacent to selected marks that may later prove the need to consider the existence of overlapping, double taps, or merging.

A discovered tendency of friction ridge patterns to exhibit axial symmetry has raised some discussion concerning the value of the characteristics located in the affected area of the marking.

2.5.2 Natural Symmetry

Symmetry has been linked to energy conservation by the discoveries of Emmy Noether. Her conclusions[13] form what is known as Noether's principle: *When a change in a physical system leaves some aspect of that system unchanged, the system possesses a corresponding symmetry.* "The existence of a symmetry means that a feature of the system is changeless, or invariant."[14]

There is a distinct possibility that some feature in the maturing fetal digit is the invariant cause of friction ridge symmetry. Whether this adds to or detracts from the evidential value of the resultant area of symmetry is a matter of debate. It should be noted that the tendency of ridges to form in a symmetrical pattern about an axis is neither a common phenomenon nor a rarity. Within that area of perceived symmetry, the location, size, and shape of the ridge unit components that make up the features do not possess a corresponding symmetry.

In most natural systems, symmetry is found in three forms: asymmetry, axial symmetry, and radial symmetry. Asymmetry is certainly present in dissociated friction ridge patterns. These do not conform to the usual arrangement of sequential ridges, let alone sequential placement of characteristics within a pattern. Axial symmetry is a recent discovery (also referred to as "mirroring"). This leaves the possibility of radial symmetry, which, although not yet defined and which I only recently posited, is considered worth considering.

A typical description of the human form is a "bilateral" organism in which the right and left halves at front and back tend to mirror each other. It is well known that the organs and many features do not maintain this symmetry throughout the body. One arm is slightly longer or shorter, one ear is not a replica of the other, and of course, friction ridge patterns vary from one hand, palm, or foot to another.

Symmetry exists within friction ridge patterns in some different ways:

1. For mirroring, characteristics along a single axis tend to mimic the location and general form from one side to the other.
2. Axial symmetry is also expressed in a more general way by most tented arches, arches, whorls, and double loops in that each tends to contain either deltas at corresponding or near-corresponding locations or at the least a similarity of pattern flow on each side of the pattern.
3. Radial symmetry is anticipated to exist around or associated with a point rather than a line or an axis. The geometry of a circle, for instance, is most often defined by a radial symmetry about a single point.
4. Looped patterns can also express a form of axial symmetry in that subsequent ridges will tend to mirror the paths of the innermost pattern flow. Placing a point at the apex of a looping ridge or the midpoint of a double recurve reveals symmetry of the adjacent features (see Figure 2.5).
5. Perhaps the most relevant symmetry is found in areas that are seldom present in an "inked original" or "scanned impression." This is the area of each digit that extends from the central pattern area to the juncture of the derma and the nail plate.

2.5.3 The Structure of a Digit

The ontogeny (sequence of biological development) of a human finger has been studied by several contributors. Researchers investigating the prenatal development of friction ridges tend to agree in observation of the symmetry of volar pads and the stresses and tensions as at least partial causes for differential growth. While there is little doubt that friction ridges develop in unique patterns, differential growth is a primary theory regarding the as-yet-unproven individuality of one pattern compared to another.[15]

The image formed by every pattern can be generally described as a set of ridges spanning the width of the digit, nearly parallel to the distal flexion crease, extending upward in a generally up-thrusting manner, running parallel along the folds of the skin that border the sides of the nail plate, and tending to join the ridging on either side by traversing across the free edge (the portion of the nail plate that is trimmed in a manicure).

This general description of friction ridge flow applies to both the toes and the fingers. The observations imply a relationship between friction ridge formations and the fingernail. A definition of the fingernail reveals some interesting details.

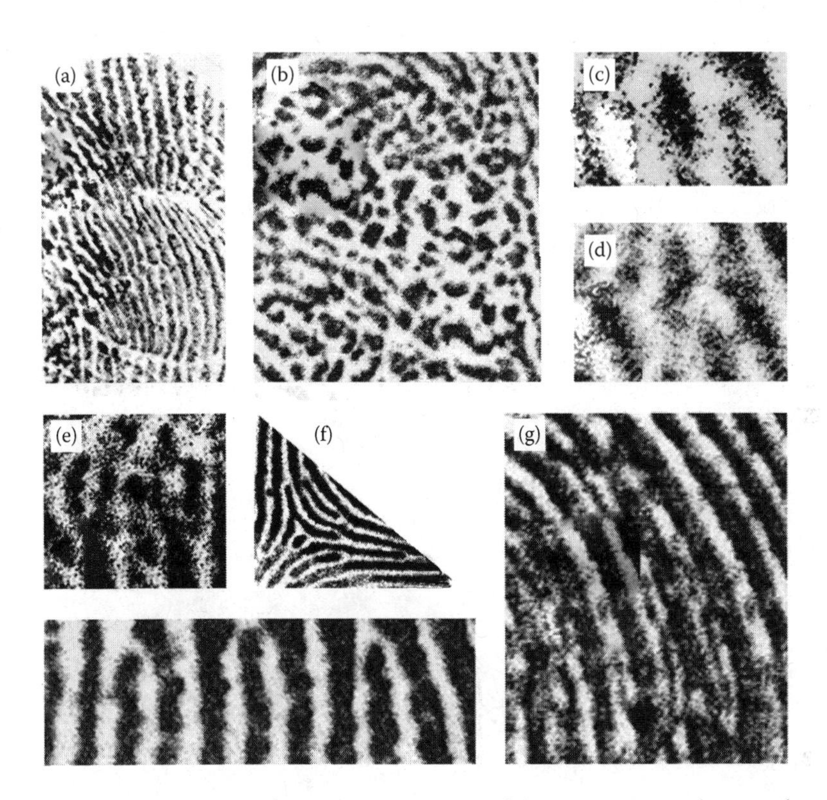

Figure 2.5 Symmetry and more. Images copied from copies to obscure details (as in pore structure) that might betray the identity of the specimens. (a) An obvious example of ridges whose paths become dissociated by injury is shown; while that is most likely the same condition for (b), these disruptions could be naturally dissociated. (c) and (d), Two examples of the most common form of mirroring in which a ridge ending is bound by two other ridge endings in axial symmetry. The two images in (e) both involve an alignment of features that may be an example of radial symmetry with characteristics that are arranged in a more or less linear way. The triangular specimen (f) is interesting in that each of the three directions leading from this delta offers some form of symmetrical arrangement. Toward the squared angle (bottom left) there are two ridge endings with two ridges intervening; looking upward, there are three features (one of them a ridge dot) that end abruptly in almost a horizontal line, and toward the bottom right there are two more ridge endings that have two continuous ridges intervening between them. (g) Example of incipient (not fully formed) ridges appearing between conventional ridge forms. These features can be mimicked in crime scene imprints by the presence of a hesitation impression, which can also result in not fully formed details between conventional ridging.

Fingernail length is sometimes used as a guide to the age of a pre-, full-, or postterm baby. The estimates are based on observations of embryonic nail growth in the ninth week. Nail growth in an adult is generally accepted as about 6 months for a fingernail and 18 months for a toenail.

Nail cells are generated within a nail root located in the matrix and emerging from under a flap of skin, known as the cuticle or *eponychium* as a hard plate that is a modification of the stratum corneum. The cells produced become heavily keratinized. The layers of nail plate, arranged in parallel ridges on the underside, are interlocked with corresponding ridges in the nail bed. "Beneath the free edge a thick ridge of skin, called the *hypoychium*, binds the nail plate to the tip of the finger (or toe)"[14(pp66–67)] (see Figure 2.6).

By comparison, the symmetry of impressions derived from manufactured items is often much easier to take into account. In the example of an injection-molded specimen, features or damage to a mold, temperatures, or

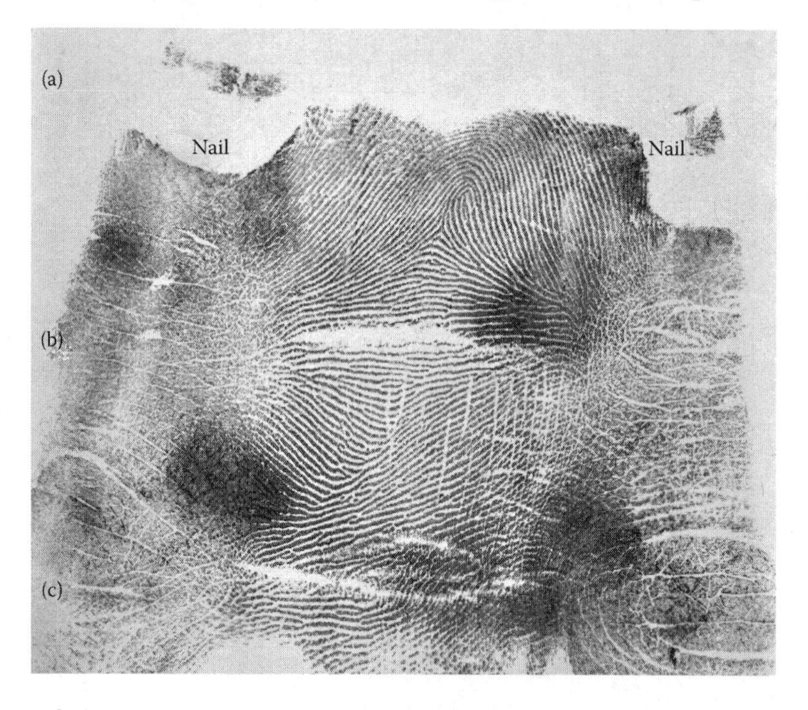

Figure 2.6 Structure of skin around the circumference of a digit. The area of a digit included in a typical inked fingerprint does not illustrate the structure of the skin. There are features that some practitioners have attributed to a particular finger (e.g., the divergence of ridges high on the digit to one side or the other) that are common to both sides of each and every digit examined in research with more than 100 individuals. The common features of pattern flow seen here are (a) the migration of ridges from a roughly horizontal condition near the distal flexion crease, to (b) running nearly parallel along the nail bed, and often joining or (c) approaching a joined condition of those parallel ridges in an arch across the fingertip nearest the emerging nail plate.

the effects of a particular design can cause repetition of the anomaly in subsequent products from the same mold or under the same conditions. Caution in attributing such features as either random or manufactured characteristics is wise.

All of those features that are found in impressions, whether of natural or synthetic original, are not necessarily dependent on just the interaction of two surfaces. There are mechanical laws, electromagnetic theories, and strong and weak forces to be considered. These are the considerations that form an area of forensic concern that has barely been examined.

2.6 Perspective

2.6.1 Offering a Different View

Imagine having the ability to move freely within the space you now occupy. Think of what your vision of the world would be if you could see your environment macroscopically, as if you were "a fly on the wall." These are the views that forensic practitioners must interpret. It is the world of fine markings and detailed contacts.

Now, take the thought a step further and think of the same materials at a microscopic or a cellular level or beyond. How different that view would be. These are visualizations that can unlock the mysteries of how surfaces behave and markings are affected; it is the realm of largely overlooked evidence.

Your surroundings at this minute consist of a sea of electron clouds that behave in as-yet-undefined ways. The electrons in that sea may travel at different rates or have looser or tighter bonds. The atoms or molecules to which they are bound are defined by those electron clouds. Each contact between one surface and another has the potential to alter the topography of the clouds in contact.

Temporary effects occur between substances that belong to the triboelectric series, in which areas of charge can be temporarily changed. Again, small forces, like the influence of surface contaminants, account for the half-life of those changes. It is also worth considering the void, vacancy, or hole left on a surface at a dipole moment. These holes occur where electrons have been attracted away from a location to which they will return when the attraction has dissipated.

2.6.2 Approaching a Different View

DNA has softened many of the boundaries between chemistry and biology. It is much more difficult in modern studies to sort out where one discipline begins and the other ends. It seems that something similar needs to occur within forensic examinations. The recognition of the strengths and similarities of all forms of impression evidence may result in the adoption of a different language. This has at least the potential to change our view of how markings are made, how they are deformed, and what information may be culled from the study.

References

1. National Academy of Science, Committee on Identifying the Needs of the Forensic Sciences Community. 2009. *Strengthening forensic science in the United States: A path forward.* National Research Council, the National Academies Press, Washington, DC.
2. National Academy of Science, Committee on Identifying the Needs of the Forensic Sciences Community. 2009. *Strengthening forensic science in the United States: A path forward,* draft corrected April 7, 2009. http://lst.law.asu.edu/FS09/pdfs/NIFS_Legislative_Outline.pdf.
3. Jackson, G., S. Jones, G. Booth, C. Champod, and I. W. Evett. 2006. The nature of forensic science opinion: A possible framework to guide thinking and practice in investigations and in court proceedings, *Science and Justice,* 46(1), 33–44.
4. Schiffer, B., and C. Champod. 2008. Judicial error and forensic science. In C. R. Huff, R. Huff, and M. Killias (Eds.), *Wrongful conviction: International perspectives on miscarriages of justice.* Temple University Press, Philadelphia.
5. Pasteur, Louis. September 11, 1872. Address In *Comptes rendus des travaux du Congrès viticole et séricole de Lyon.* September 9–14, 1872.
6. Dawkins, R. 2008. *The Oxford book of modern science writing.* Oxford University Press, New York.
7. Brown, T. L., H. E. LeMay Jr., B. E. Bursten, and contributing author Catherine J. Murphy. 2006. *Chemistry: The central science.* Pearson Education, Lawrenceville, NJ.
8. Houwink, R., and H. K. de Decker. 1971. *Elasticity, plasticity and structure of matter.* 3rd ed. Cambridge University Press, New York.
9. Paredez, A. R., S. Persson, D. W. Erhardt, and C. R. Somerville. June 26, 2008. Cellulose synthase influences microtubule organization. *Plant Physiology Preview.* http://www.plantphysiol.org.cgi/rapidpdf/pp.108.120196v1.pdf (accessed June 25, 2010).
10. Ioelovich, M. 2008. Cellulose as a nanostructured polymer: A short review. *BioResouces,* 3(4). http://www.ncsu.edu/bioresources/BioRes_03/BioRes_03_4_1403_Ioelovich_Cellulose_Nanostruc_Polymer_Review.pdf (accessed June 25, 2010).
11. Shafiei, A., and T. Alpas. 2008. Fabrication of biotextured nanocrystalline nickel films for the reduction and control of friction. University of Windsor, Materials Science and Engineering, C 28. http://www.sciencedirect.com/science?_ob=ArticleListURL&_method=list&_ArticleListID=1381999836&_sort=r&view=c&_acct=C000050221&_version=1&_urlVersion=0&_userid=10&md5=de1078c832cf1015ea09fd3ced0cc484 (accessed June 25, 2010).

12. Pierce, D. S., and S. S. Turnidge. 2008. The significance of butterflies. *Journal of Forensic Identification,* 58(6), 696–711.
13. Hecht, E. 1999. *Physics: algebra/trig.* Second ed. Brooks/Cole Publishing Company, Florence, KY.
14. Stein, L. 2007. *Body, the complete human.* National Geographic, Washington, DC.
15. Swofford, H. J. 2008. The ontogeny of the friction ridge: A unified explanation of epidermal ridge development with descriptive detail of individuality. *Journal of Forensic Identification*, 58(6), 682–695.

Ivory Tower Syndrome 3

A mathematician confided
that a Möbius band is one sided.
And you'll get quite a laugh
if you cut it in half,
for it stays in one piece when divided.

—Unknown[1]

3.1 The View Has Changed

Like a Möbius strip, an "ivory tower" may be replicated rather than expunged depending on the method of separating engrained habit from practice. The transition to science from deductive habit is complicated by the "lack of fit" between a reasonable conclusion and one that can be supported by statistics. There has been no suitable application of numerical comparison that encompasses the evidence seen in an amount of detail, and until there is a bridge of that sort, there is the danger of supplanting one ivory tower with another.

3.1.1 The Making of Towers

The world in which we live is a collective environment filled with interdependencies. This connection may not be readily apparent, and without a wider perspective, one can easily become comfortably ensconced in an ivory tower of his or her own making. Forensic disciplines are on the verge of emerging from just such an artifice.

Looking to lessons of the past is a good way to learn about what is worth preserving. Critics of the current state of forensic science are all too concerned with those things that are deemed unscientific. That is no reason to discard the progress that has been made or the experience developed or to ignore indications of the pitfalls that may well populate the future of forensic science.

The concepts that "every contact leaves a trace" and "no two things in nature are exactly alike" are corruptions of what was pointed out by scientists from times past. These tenets, quoted correctly or not, have been part of the fabric of an ivory tower. It must be noted that while many practitioners and scientists agree that a more methodical approach should be adopted to replace dogma, the accuracy of deductive reasoning was never disproven.

Science, religion, culture, industry, ecology, and philosophy, to mention a few, connect in one way or another to form a definition of their common environment. It would be easy to assume that science provides an alternative to secular concerns, but science is also populated with a wide variety of ivory towers. Many of the sciences depend on theories to explain a thing or a phenomenon that require another theory to prove the first theory, which is fine until a supporting theory changes.

Each discovery about the atomic nature of matter caused speculation that there could be no smaller unit. Theories about matter have often changed, and current models are tempered with caution. The "solar system" model of the atom may be useful in some circumstances, but you should know that the electron is much more unusual than that model suggests. The electron is extremely tiny, and modern physics tells us that "strange things happen in the realm of the very, very small."[2(p454)]

The activities of Pythagoras and his followers[3] are not well recorded. What is known of the school is that they apparently attempted to explain their environment (including science, religion, and philosophy) in various forms of mathematics. Today, this group, including their curious vegetarian diet that excluded beans, would be labeled a cult (a form of ivory tower) in honor of their eccentricities.

The methods of justifying a conclusion are at the heart of a forensic ivory tower; deductive and inductive reasoning require a leap of faith to be considered useful. Deductive reasoning, previously regarded as a cornerstone of forensic justification, represents the position that if a thing has occurred reliably over a great period and great number of times, then it will, in all likelihood, continue to be as reliable in the future as it was in the past. Inductive reasoning, by comparison, takes hold of the same problem, analyzes examples, and relies on a statistical likelihood, which can never reach a certainty, that the same event will be as reliable in the future as it was in the past.

Bertrand Russell[4] introduced an apt analogy that pertains to the justifications for reasoning. If you consider that animals, humans included, become accustomed to a particular recurring event, then there is a level of expectation that arises from that routine norm. A chicken, for instance, will expect food from a farmer who has provided food each day for the chicken. It is safe to assume that the chicken does not have any reason to expect anything other than food on the day that the farmer wrings its neck.

The chicken argument was directed at deductive reasoning, even though it would not matter to the chicken if it were also applied to inductive reasoning. This oversimplification points out the chief advantage of inductive reasoning: When the unlikely occurs, inductive reasoning can fall back on its error rate. This is an advantage that would not impress the chicken, has no effect on the outcome, and only serves to placate a legal system that wants a plausible escape from the possibility of a wrongful conviction.

An expert who offers the opinion of a high degree of probability (rather than a deduced inference) concerning a conclusion has not offered any additional evidence. The layer of statistical probability merely offers a security blanket of sorts. The practical advantages of reconsidering the way in which things are achieved in forensics lies in addressing fundamental issues of bias, implementing validity in each phase of the methodology, and infusing a continuance of education.

Certification, one of the recommended solutions to the problem of training, offers its own flaws. Designed to be used as a standard of practical knowledge, certification will eliminate only the most obvious issues of performance. Certification is a minimum standard, not a guide to, measure of, or incentive for improving the actual proficiency of the person giving evidence.

Certification could be considered an allowable ivory tower in that you have to start somewhere. A general test (like certification) can be fitted to a legal balance as the tare point at which the judgment of the competency of a witness begins. Training and achievement past certification then become a more meaningful gauge of expertise.

The role of the expert is, as it ever was, to provide the best possible evidence in any matter, as though acting as the eyes of the court. It is not at all practical to expect most forensic witnesses to be able to offer useful opinions of statistical probability, but it is wise for such a witness to have achieved not only a thorough understanding of the topic at hand but also a demonstrable level of proficiency above and beyond minimum standards.

Attending many scenes, receiving training, or even passing an examination regarding a course of study are useful activities. These offer a lesser pedigree of expertise than participating in research, particularly if those results are shared by means of educational sessions, lectures, or publications. Too often, administrators stress the value of numerical or statistical evaluation over the quality of what has been or could be achieved.

3.1.2 Descending a Tower

The preoccupation of a demanding profession tends to narrow one's view regarding the significance of achievements. Those that covet acceptance of their own advancement over the greater good tend to create kingdoms where secular views can thrive. Greater minds expect differences, encourage exploration, and promote advancement through shared research.

The various forensic disciplines have responded quickly to criticism, and one hopes that there will be time and a mechanism to accurately evaluate the quality and direction of this evolution. Science is not without its changes; a recent example is the modeling of the structure of an atom, including an electron cloud, which replaced the previously considered strictly orbital configuration.

The introduction to this text suggested that there are times when impression evidence can be expected to provide greater value than DNA. Proponents of DNA as the only "true" form of evidence would rail at the suggestion, as much as the practitioners of other disciplines would have reacted to the initial suggestion that their discipline lacked scientific justification.

There have been attempts at applying some numerically based justifications to footwear comparisons. Various groups, North American and European in particular, have adopted standardized language intended to circumvent the problem, but the arguments have not established a global common ground, and that is why rethinking is needed. There needs to be agreement that creates a more unified and elemental basis, including language, protocols, and procedures that transcend the borders of nations or continents, one that encompasses the mechanics concerned with each comparison.

Imagine the effects of challenging the notion that parallel lines can intersect (as proven by non-Euclidean geometry). One may be so comfortably enshrouded by a belief that they may not even realize that it is no more than an ivory tower. The onslaught of a well-placed bit of unfamiliar logic may topple the tallest ramparts.

The ivory towers associated with forensic science are under siege, and the current focus is the lack of reliability cited by scientists and scientific organizations such as the American Academy of Science. The premise of the attack is that forensic evidence is not scientific. This, in their opinion, is demonstrated by the lack of empirical evidence, testing, protocols, application of the scientific method, and a lack of training and certification of personnel. The criticisms are warranted to some degree, yet they are also the foundations of a new set of ivory towers.

A collection of principles used as the basis for any endeavor or belief tends to become assimilated as if part of the fabric of the process. Acceptance of the doctrine leads to perceived security and often widespread complacency in the absence of challenges. Any questioning of a founding notion will commonly elicit surprise and disbelief.

Advocates claim that empirical studies justified with statistical analysis make up the only truly reliable method of providing evidence and hold forth DNA as the gold standard. This approach works well for DNA analysis in which the component of human interpretation can be minimized with relative ease. One day, there may be computer programs that allow for more comprehensive forms of machine analysis for all forensic endeavors, but this technology has not yet been perfected. The interpretation of a distorted impression is an example of something that cannot be achieved by current technology in all but the simplest of terms. Perhaps the popular quotation attributed to Albert Einstein (1879–1955) phrased it best: "As far as the propositions of mathematics refer to reality they are not certain, and so far as they are certain, they do not refer to reality."[5(p28)]

3.1.3 The View at Ground Level

The arguments favoring uniqueness theories stem from classic philosophers such as Plato and more recently the theories of Gottfried Wilhelm Leibniz (1646–1716), who postulated in a metaphysical sense that if two entities were indistinguishable, one from the other, they are one and the same entity or, in other words, that an object can be *identical* only to itself (Leibniz's law).

Study of the history of science shows that methods of discovery consist of more than a basis rooted in empirical proofs. R. L. Gregory, in describing the contribution of Greek society to science stated: "Thales was a great inventor and engineer, and was a primary figure in introducing mechanical models into speculative thinking, using actual or imaginary models based on the technology of his day for explanations by analogy, which remains the basis of science."[6(pp21–22)]

Gregory further noted that "Aristotle founded Deductive logic, by formulating syllogistic arguments, and also (which is sometimes forgotten) he promoted Inductive procedures for science." This attribution is accompanied by the following realistic caution: "Aristotle's notions of mechanics seem intuitively right before they are questioned by critical experiments, and children today seem to think as he did."[7]

One of the most significant developments in the history of forensic science and traditionally the event that marks the transition from classical to modern theories occurred in December 1888, when Francis Galton (1822–1911), an English investigator into heredity and cousin of Charles Darwin, published his paper, "Co-relations and Their Measurement, Chiefly from Anthropometric Data."[8]

Dramatic events such as challenges that attack the cornerstones of a profession tend to elicit a defensive response. The urge to protect the integrity of a particular subgroup was the likely motivation for a particular forensic expert to describe friction ridges as consisting of only three characteristics: the bifurcation, ridge ending, and ridge dot.

Circles, squares, and triangles form the basis of all three-dimensional shapes, yet to say that the impression of a tire track, footwear imprint, or tool mark is nothing more than shapes would be misleading. Simplification is a great tool, as is routinely used in mathematics to make complex matters more workable, and is similarly useful in explaining the greater complexities of forensic analysis. Simplification is necessary to provide explanations in court.

An ivory tower is created when one loses or never gains a perspective regarding the relevance of tangents and the ties between related endeavors or professions. This lack of vision becomes an unrecognized disadvantage.

The various forensic disciplines all engage in separating and justifying classes of characteristics to achieve some degree of correspondence or incongruence between impressions and the suspected source. This is the nature of evidence, no matter the source or type and no matter if the information incriminates or exonerates:[9]

> The use of earth materials, soils, rocks, minerals, and fossils, as for all physical evidence, has both limitations and advantages. The fundamental limitation lies in the fact that significance of such evidence is determined by probability and statistics. No two physical objects are ever exactly the same in a purely theoretical sense. For example, if we were to take a rock from any outcrop and break it into two pieces, in most cases it would be possible to show, by detailed study, differences between the two pieces. The similarity between the two pieces in most cases would be large and we would be able to say that the two pieces compare, and that there is a high probability that one piece was a sample of the other. If the two pieces could be fitted together and individual minerals could be shown to be broken and lined up when fitted back together, then the probability that they were once part of the same rock would be even greater. In this case we would say that we had shown an individual characteristic and that there was really no doubt of the comparison. We would, however, still be dealing with probability and the value of the determination would largely depend on the competence of the scientist who made the determination and availability of data.
>
> In many cases involving earth materials, the probability becomes so high as to approach that of the individual type of evidence such as fingerprints. For example, in a Canadian rape case the knees of the suspect's trousers contained encrusted soil samples. The sample from the right knee was different from that collected from the left knee. In examining the crime scene, two knee impressions were found in the soil corresponding to a right and left knee. Samples taken from these two knee impressions were different. The soil sample collected from the left knee impression compared with that removed from the left trouser knee of the suspect as did the right knee impression and the right trouser knee. A major change in soil type occurred between the two knee impressions, indicating that a contact between two rock or soil types was located at that place. (pp. 24–25)

This quotation refers to the status of acceptance for evidence before the advent of Daubert challenges. The world before 2003 was an example of an ivory tower of sorts that was toppled in short order. The other "subgroups" of forensic science were not long in experiencing the fallout. Forensic science will evolve; it has to. The change will affect lab staff and practitioners alike. The role of "expert" will see the most upheaval.

There are principles of identification; they relate to the unique sequential attributes of both markings and the entire impression. These principles, at first glance, would appear redundant in light of empirical methods. In essence, the scientific method substitutes a statement of probability for the more encompassing deductive premise of individuality. Statistical probabilities are not all that far removed from a premise of individuality since each is nothing more than an extrapolation based on axioms.

Scientific evaluation is, it appears, being adopted as a new ivory tower. Adoption of a statistical basis is a reasonable option only if the pursuit of measurement does not corrupt the process it is intended to validate. The biological nature of friction ridge units and associated details does not fit as easily into a statistical framework as do the other inanimate donors of impression evidence.

Few things would be easier for "fingerprint experts" than acquiescing to the pressures of Daubert challenges with a reply that offered some ridiculously high probability of mutually exclusive individuality. The fact is that no ironclad probability algorithm for friction ridge individuality exists, and it is unlikely that there will ever be one.

Forensic examiners have been as reluctant to surrender their ivory tower based on deductive reasoning as scientists are adamant that empirical techniques create the only reliable form of evidence. Do not be confused by this statement; empirical methods are both necessary and useful. The fact is that to blindly accept empiricism is to reject the spirit of science:[10]

> After all, the accounts of science do change; indeed they change far more than do the myths of antiquity! How can we trust the ever-changing assertions of science as true? Here I suppose that we may accept that science is homing in on truth, and indeed, few scientists suppose that we have arrived at a fixed account, though there have been periods, especially in the nineteenth century, when this was believed. (p. 35)

Many well-known scientists were not university trained and achieved serendipitous discoveries that may not have been found by conventionally trained individuals. Two examples are Thomas Alva Edison and Percy LeBaron Spencer; both patented more than 120 inventions, and neither had finished grammar school. Spencer had been working with microwaves in the mid-1940s and formulated the hypothesis that the energy waves had caused a candy bar in his pocket to melt. "To test his notion, he exposed a bag of popcorn kernels to the microwaves."[11(p10)]

Science affords one a basis on which to make predictions. There are some predictions about the infusion of more science into the disciplines of forensic science. The first is that applications of science are needed throughout the forensic disciplines; second, these applications will lead to further examples of ivory tower reasoning.

3.2 The Basis

One must strip away the ivory towers to lay bare the basis of the discipline. This differs from the fundamentals of a technique in that a basis provides the means to make use of the fundamentals.

basis /'beisis/n. (*pl.* **bases** /-si:z/ **1** the foundation or support of something, esp. an idea or argument. **2** the main or determining principle or ingredient *(on a purely friendly basis).* **3** the starting point for a discussion etc. [Latin from Greek, - BASE']. (*Oxford Concise Dictionary,* 9th ed., 1995)

3.2.1 An Ability to Perceive

Machines such as cameras or scanners are specialized pieces of equipment that operate predictably. Their optimized method of operation is independent of the subject matter. Machines that "read" or scan friction ridges present an array of scanned portions that are blended to achieve remarkably consistent results, which does not absolve machines from the possibility of error but merely attests to their objectivity.

This is not how the human eye functions. "What the eyes do is to feed the brain with information coded into neural activity—chains of electrical impulses—which by their code and the patterns of brain activity, represent objects."[12(p7)] Technology will need to leap into some new realms before the human eye can be replaced as the sensor of choice in the comparison of fine details.

Challenged by a complex pattern, our thought processes want to make use of similarity (a starting point) between corresponding portions of two images. The methodology of advancing from a starting point to a progressive comparison of characteristics in sequence will often vary from one examiner to another. It is not uncommon, if there are differing levels of expertise or complexities in impressions, that one can find different paths to the same conclusion. It is most often the nature of the impression (if a holistic view includes elements of distortion) that will affect the choice of a path for comparison. This can render any recommendations of a mathematically predetermined search method unwieldy, if not useless. In advance of contemplating the psychology of this task, we need functioning knowledge of the mechanisms of the eye.

Without light, there is no sight. Our range of vision is restricted to a small portion of the electromagnetic spectrum. The world of visible electromagnetic radiation can also be defined as light emanating from a point source.

Light exists in particles (photons) that travel in electromagnetic waves. Auditory signals travel in sound waves, olfactory signals are carried by air currents, and tactile impressions need contact. Each sense requires a medium, is affected on an atomic level, and provides some relative amount of evidence. A consideration of the mechanics of evidence must relate to how information is detected and perceived.

Lighting affects our senses, perception, and health. Lighting can vary: a "Rembrandt" style as used in portraiture, the "mood lighting" used in interior design, and the task lighting of a workstation. Ultraviolet light can be harmful, but like plants, we need sunlight to thrive.

The effects of varied lighting can work on a subliminal level or cause immediate changes in the way we perceive the objects around us. A simple experiment with colored filters over area lights can make it impossible to detect butter from cheese. The angle of lighting is well known to affect our impression of whether an impression is indented or protruding from a substrate; this is known as the "inversion effect."[6]

No object or situation can become evidence without setting into play a sequence of events that begins with perception. This function is dependent on historic examples within memory and the ability to make use of a sense. A sensory signal, taken alone, is meaningless without an associated memory. The combination of sensory input and successful association may then be considered an impression, theory, or conclusion, depending on the strength of certainty attached to the experience.

If that were all it took, there would be no need for further analysis. We would be able to solve every question at a glance, touch, smell, or sound. The senses operate in nonconforming ways. The tactile sense of touch differs in sensitivity from one part of the body to another. The location of taste buds on the tongue, differences in hearing or hearing loss, and sensitivity to odors can differ with time, and all demonstrate the need for caution if the association of one thing to another is intended as a source of evidence.

Vision is generally considered the most discerning of the senses. The eye operates on two levels; foveal vision is detailed but slowly processed, and peripheral vision is both quick and low quality.

Independent of image or sensory quality, there is still the question of what the eye does with those perceptions.

Initiating the process of association, a pattern of light is received through the lens of the eye upside down or transmitted as different signals by nerves from the fingers, ears, or even the nose. Those transmitted sensations are then associated within the brain. "Indeed, we may say that perception of an object *is* an hypothesis, suggested and tested by the sensory data."[13(p21–22)] This is obviously not a purely empirical procedure; it involves aesthetics and is subject to the experiences of the individual.

The fact that patterns from the senses are interpreted in an associative way by the brain leads us to examine the subject of structural pattern recognition. This is the essence of language.

3.2.2 The Ability to Describe Basic Properties of a Material

The study of impressions is a study in material science that requires a description of the result of a contact. To describe the product effectively, there needs to be some grounding in the basic language of materials and material interactions. This section begins with a brief description of matter that encompasses part of the definition of substances, definition of simple properties, and an introduction to material behavior that goes beyond fundamental concepts by showing how these facts can be integrated into a description of the basis from which to describe other materials.

The simplest forms of description are merely observations of the apparent properties of materials that allow one to describe a pure substance from an aggregate or an aggregate from an agglomeration (such as a cement or laminate). These descriptions will almost routinely progress with descriptions that become entangled in measurements of various properties from solvencies to particulate size.

The description of mechanical properties includes characteristics such as stress and strain. In solids, this is an initially proportionate relationship, as strain disappears with the removal of a load (stress). These are conditions within the elastic limit of a substance known as Young's modulus of elasticity, $Y = \sigma/\varepsilon$, measured in units of newtons per square meter.

We are all familiar with the visible properties of materials, but forensic testimony regarding materials often lacks the language that binds findings to other sciences. We may know that steel is stronger than copper, but how is this known? And, what is the relationship of other common materials?

Comparatively, steel (21×10^{10} n/m^2) is about one-half as likely to deform as copper (11×10^{10} n/m^2). Window glass and aluminum (7×10^{10} n/m^2) exhibit roughly similar resistance to deformation (at about 2/3 that of copper). Polyethylene (14×10^8 n/m^2), at its strongest, possesses about 1/50 the resistance of window glass or aluminum, and rubber (80×10^6 n/m^2) at its most resistant has about 1/10 less resistance to deformation than polyethylene. These descriptions apply as a portion of the basis on which materials may be compared.

Those figures describe only the elastic properties of stress and strain. The point at which elasticity is exceeded is the breaking strength of the material. That is the point at which atoms of the material will no longer return to the same location relative to neighboring atoms and will instead relocate to a new position relative to the amount of fracture. This is known as plastic deformation, which helps to explain the wisdom of choosing the term plastic impression to describe a (relatively permanent) three-dimensional impression.

The description of hardness for solids such as metals is also a description of the ability of a substance to resist a force applied to it. With metals, a Brinell hardness number (BHN) provides a comparative value based on the amount of pressure required to cause an indentation in the surface of a specimen. While the diameter of an indentation can be expected to provide a correlation to the hardness of the substance, these results are erratic with brittle materials, which may be affected by flaws such as stress fractures that negate any attempts to provide good correlations of hardness and strength with those materials.

The generally accepted definition of toughness applies to a material that is strong in that it can withstand high stress and exhibits considerable ductility (able to withstand great strain before fracture). Measurement of toughness is expressed in joules per cubic centimeter, which provides a comparative evaluation of the amount of energy required to cause a fracture in the form of $E = fd$. A pendulum tester can be used to provide such calculations (see Chapter 10).

The effects of various properties can also be compounded to alter what we would expect to see. A rubber tire, for instance. will exhibit great elasticity in response to an application of force within the boundaries defined by the design of the tire (e.g., winter tires are made softer for improved grip). A 9-mm bullet passing through a deflated, partially inflated, or fully inflated inner tube will cause little stretching and leave an easily measured aperture in the specimen. The velocity of the round allows insufficient time for the rubber molecules to react to the stress, thereby localizing its effects.

In evaluating aggregates such as soils and clay, a compressibility figure is similarly the measure of a response to an applied force. In that instance, the water content variability must be controlled to arrive at useful comparisons. Successively drying a specimen in a kiln and weighing the specimen after cooling until there is no difference from one result to another ensures that the moisture content has been removed. This becomes a useful starting point for controlling the reintroduction of moisture content and other additives.

Thermal properties can also help to define a substance. Thermal expansion can explain changes in the length of a wire under tension, alter the volume of substances at proportionately different rates, or crack open rocks when combined with freezing. Higher temperatures introduce vibration and some disorder among atoms, even within a solid.

The amplitude of thermal vibration effects is considered in two ways; a volume expansion coefficient and a linear expansion coefficient offer a generally accepted reference point for comparison of thermal properties between one material and another.

At approximately room temperature, building bricks and window glass have a similar linear thermal expansion coefficient of about 9; iron and concrete have relatively similar values at about 12 and 13, respectively; and values are 81 for rubber and 110–180 for polyethylene (all values are times 10^{-6} per °C). The values for other metals range from 16×10^{-6} for copper to 22×10^{-6} for aluminum. The last figures help to explain the problems associated with the now-discontinued introduction of cheaper aluminum residential electrical wiring between 1965 and 1973.

Aluminum wiring exhibited some properties considered to be deficiencies. Cold flow (or creep, the slow deformation of material under stress), brittleness, oxidation, and a high coefficient of expansion caused concern about electrical connections with this wiring, leading to a number of guidelines for maintenance or repair. In regard to the coefficient of expansion, aluminum has been cited as having the capability of expanding 30% more than copper under heat stress.[14]

3.2.3 The Changing Basis that Begins with Reason

There are three types of reasoning that affect impression evidence:

1. Deductive reasoning: A premise is used as the "gold standard" by which to compare and it is hoped arrive at a similarly unquestionable conclusion.
2. Inductive reasoning: An isolated question is subjected to a series of tests or proofs that must then be evaluated regarding their accuracy (or error rate).
3. Abductive reasoning: A plausible answer to a question is tested to see if it remains the most likely explanation.

Reasoning is a part of every decision we make; from habits to choices, we reason our way through life. Effective reasoning has an obvious impact on our successes. To understand the use of the types of reasoning, we can start with everyday models. We tend to be most familiar and comfortable with *abductive* and *deductive reasoning* since these two methods are more productive in a practical sense.

To understand the mechanics of the impact that reasoning has, look first at the issue of what to wear on a given day. Deductive and abductive reasoning are similar concerns in this respect; you either operate on the premise that a certain time of year requires a different comfort level or develop a personal theory regarding the most likely expectation. You may look for proofs such as checking a weather report or observing phenomena, but your decisions will be derived from a combination of premises, theories, experience, and clues.

There are reasons that the scientific method does not apply to everyday situations. Direct data about weather can only be obtained by being where the weather is occurring. While the results would be precise for the moment, they are not flexible enough to accommodate "reasonable expectations." Once outside, you might take a number of readings, and on returning to the indoors, you would need to calculate the probabilities, including an error rate.

Imagine the scope of clothing choice dictated by improving the error rate. Snowshoes in summer anyone? Even tempered by probabilities, empirical methods are most useful only when precise observation and calculations are required from relatively confined situations.

This is not the type of reasoning that allows for quick or intuitive results that are needed to process a crime scene. In a lab setting, there is more time but not enough to allow the relentless probing posited by empirical methods. The choices of what to examine and the order of priorities must be dictated by the circumstances; these can be a matter of life and death.

Acceptance of a theory within scientific circles is a long process. Scientists mistrust those thoughts that have not suffered the burden of multiple proofs. New ideas are not generally welcomed without benefit of scrutiny from various angles.

Describing a degree of probability that the sun will rise tomorrow morning is neither a statement nor a question as much as it is an argument. The negative proposition has no basis in common experience or in historical context, yet the possibility must be conceded that there may be an occasion when the sun will not rise as expected. The task of calculating the probability seems particularly futile and impractical.

The list of prominent individuals having given serious consideration to the matters of evidence, proof, and justification is impressive. The list could easily extend from antiquity to the present. It is plainly not a sensible undertaking to attempt categorization of the various results of such deliberations within this text.

There are, however, a few matters that seem to possess enduring relevance. First, consider the nature of physical evidence. Paul Kirk suggested that such evidence includes fingerprints, footprints, fibers, tool marks, blood, semen, and more as bound to the principle that "physical evidence cannot be wrong."[15] This simple statement, and corruptions of it, forms a cornerstone of modern forensics since the evidence cannot be blamed for any errors made by those that misinterpret what they see.

The line of thought culminates in a proposition that indicates the vulnerability of humans to err. Calculations regarding the probability of the existence of error relate to the actions and processes of the witness more than the miniscule possibility that two objects, individuals, or situations exist with precise exactitude. Whether the probability of an individual thing can be found to repeat naturally with a certainty of 95% as compared to 99% merely offers the appearance of safety in the apparently greater confidence level rather than a tangible form of evidence.

Suppose for a moment that two fingerprints (friction ridge impressions) offered precisely the same information over an impressive number of specifically oriented characteristics that corresponded in size and shape, even down to the ridge units and pore locations. Further suppose that the correspondence exceeded, say, 10 such characteristics. That would mean two individuals were found to exist who exhibited exact correspondence over the course of a tremendous amount of detail.

The logical questions posed by these thoughts are as follows:

- What can be agreed regarding a level of correspondence that would be simply insufficient to support a proposition of evidential value?
- If evidence is more or less relevant based on a numerical assessment, what are the demarcation points for the weight of the evidence?
- What number of exceptions to opinions of individualization is required to diminish the evidential value and constitute unreliability in evidence pertaining to identity, such as friction ridge identification, tire tracks, or DNA comparisons?

It is fortunate to note that the answers are not yet required; they will only become entirely relevant in the unlikely event of a blatant failure of a particular discipline to provide individualizing evidence. It is also unfortunate that we can never altogether purge doubt but only hope to contain and restrain the effects of human error. This then introduces the concept of expert procedures.

3.2.4 The Identification Process

That identification or individualization is a process that requires great attention and careful consideration is not in doubt. The level of understanding of the process can and does vary greatly according to the experience and confidence of those considering it. Remember that a voiced message has little effect if not received.

The legal community, judiciaries, and juries must be sufficiently prepared to evaluate evidence as it is delivered. Scientists have historically proven the inability of complicated testimony to sway opinions. The truth of their testimony has little bearing on the ability of others to grasp that truth.

The identification process is not a new topic, but it is reasonable to consider this an essential area of study. Simplification of the process to the acronym ACE-V (analysis, comparison, evaluation, and verification) has achieved wide acceptance. Simplification, however, comes with a price, and even the credibility of methods requires further justification.

A single impression usually occurs as an assemblage of component markings that appear in a recognizable sequence. Each discernible portion (even voids) within an impression relates to a source. The source will be a specific portion of a donor, the substrate, or an inclusion.

Tracing a marking to the source becomes a creation theory that provides the first crucial step in determining the evidential value of the overall impression. Recording and accounting for each step and observation of the nature, general appearance, distortion, or interference must take the form of both written and visual (imaged) details.

Every practitioner is taught to expect trace evidence as the product of an exchange. What is not covered in most training is just how many cases possess evidence that is overlooked, underevaluated, or ill managed. Recognition and understanding require an appreciation of the significance in all forms of evidence and the mechanics of creation theory that allowed those contacts to exist.

In a "raw state," the evidence in question may appear to be visible, invisible (latent), or in flux between the two. In addition, the status of an impression may be transitory. Evaporation, contamination, or damaging effects can all impede the life expectancy of an impression. Grass that is laid flat by vehicle tracks will not be visible long; the same is true for footwear imprints on carpet, tool marks impressed on a sponge, or friction ridges on a donut. These examples all illustrate one aspect of the influence of a substrate on the longevity of impressions.

Records of evidence need to facilitate application of "principles of identification."[14] Principles need to be defined for any process (e.g., analysis, comparison, evaluation, and verification) for which the ownership of a marking is in question. That there are principles and not just methodology needs acknowledgement no matter the exact phrasing. The following principles are essential:

1. The evidential value of an impression is dependent on locating and illustrating sequential details within that impression that possess sufficient clarity and quantity to support defensible opinions of source that may vary in stature from identity to exclusion.
2. Any discrepancies can be reasonably taken into account.
3. Root causes for the appearance of an impression must be explainable, and that explanation can only be based on a firm understanding of the materials involved and the forces acting on them.
4. Findings must be reliable and verifiable, with an absence of bias.

The application of acceptable statistics in support of any findings can lend credibility and make the difference with the acceptance of the evidence if statistics can reasonably be applied. The use of statistics is much more complex with most forms of impression evidence due to the difficulty of interpretation of intrinsic details within impressions.

When technologies can be advanced to the point of making a comparison of an impression as simple as the comparison of genetic markers on a chart, then the layer of those statistics can be predicted to yield the spectacularly small chances of error currently stated in regard to DNA evidence.

Diligent record keeping is an essential ingredient of the scientifically defensible criterion that is demanded for all forensic disciplines. The overlooked elements of recording a scene are cognitive factors that include, but are not limited to, perception and intuition. Recognition of potential evidence, often guided by experience, will ultimately define the effectiveness of any search or analysis.

The record-keeping portion of the task is essential and has evolved to include records that allow the courts to see the progression of scene examination and evidence collection and preservation through comparison processes to the final presentation. Technology can be expected to alter the way in which one approaches the tasks, but the goal of accurate documentation is to provide the best possible objective portrayal of the entire process. The efficiency of the records can turn clues into evidence.

Management directed toward achieving a practical solution for processing evidence and scenes is limited by the workforce, training, equipment, and restrictive budgets. Doing more with less is always a benefit. A task that is approached with a reasonable level of competence and preparation will almost invariably be completed more efficiently and better than the blind situations that can otherwise occur. The choice of appropriate training will be partly determined by the specialized demands of the terrain and situations encountered within the boundaries of the workplace.

Removing the mystery of how a particular impression is formed, discerning the state or states in which it is found or can be expected to evolve into, and developing a common understanding of the mechanisms of formative or distorting influences is tackled in this book. The question of asserting a basis from which to evaluate those impressions is currently steeped in controversy. The simple fact on which all factions can agree is that opinion evidence regarding impressions requires a basis. While this text may appear to prefer one argument or technique over another, the fact is that the mechanisms of the bases are thought to deserve as much consideration as any particular function, observation, or conclusion that they are meant to support.

No matter which discipline is your affiliation, there are similarities within each that allow the drawing of parallels. The following description of the legal status of forensic toxicology can easily apply equally to those engaged in the study of impression evidence:

> Most commonly, it is the opinion of the forensic toxicologist that is sought, the very thing that elicits the best in the adversarial nature of attorneys. Rarely are toxilogical opinions certainties, but if done carefully and with Aristotle's three appeals in mind, i.e., Ethos (ethical appeal), Pathos (emotional appeal), and Logos (rational appeal), the process can be intellectually rewarding.[16]

The toxicologist's job is to teach the trier of fact something so that a decision can be based on all available information; it is not the toxicologist's job to be an advocate. Science by definition is truth seeking, and forensic toxicology falls smack in the middle of that definition. Further, "in this way the field will get closer and closer to what can never be attained in most cases in terms of absolutes, but, as we do today, with stronger and stronger reasonable scientific certainty than those before us."[14]

Examinations of markings must provide an accurate record of the condition, detailing not only the crime scene marking but also the substrate and any adjacent or contiguous circumstances, such as the tendency of a malleable substrate to "migrate" away from sources of pressure. The analysis of impression evidence has survived for many decades as little more than a cursory visual examination. Improved note taking is a step in the right direction.

Efficiency in observations, notes, and conclusions requires consistency of terms and a reliable basis (as supplied by the validation of techniques) on which to evaluate the worth of observations. Observations that lack communal approval are regarded with suspicion. This is a common theme among scientific communities, which often feel the impact of evolution in language even though change is inevitable.

3.2.5 Developing a Physical Basis

Every impression, no matter what its value or circumstance, begins with some form of contact. The complexity of interactions between surfaces has been a sufficient excuse in the past to avoid or dismiss the topic as though it were beyond comprehension. Contact mechanics are not out of bounds, and they may not be all that difficult to apply.

Familiarity with the Cartesian system of coordinates (graphs created with an x, y axis) is a good place to start. The system starts with a baseline x that runs positive to the right and negative to the left, with 0 as the y intersection point and with positive y values running up from 0 and negative values down. This arrangement allows plotting of a multitude of two-dimensional coordinates that, when represented by an algebraic expression, allow one to plot a straight line according to the equation $y = m \times +b$, where m is the slope of the line (rise over run), x is the run, and b is the intercept point for y.

The important thing to remember is that this simple equation can be used to plot or measure the slope of any straight line. The usefulness is apparent in graphing, estimations involving trajectories, or figuring out the difference between one slope and another. Knowing the slope of a surface is crucial information for creating test impressions that re-create the conditions under which a marking was made, and it is a much more efficient way of describing surfaces that only requires some basic mathematical ability to understand.

Using the coordinate system, it is possible to plot shapes as well as lines. Creating different figures simply requires a different equation or set of equations.

The next step is to increase your vocabulary to include the z axis, which could be thought of as a line perpendicular to both the x and y axes so that it comes off the page from point 0, giving any equation a three-dimensional reference. This model now consists of a plane defined by the increments of an *x*, *y* coordinate system that includes a *z* axis that is perpendicular to both *x* and *y*. This orientation of the *z* axis is also referred to as geometrically square to the axis.

Obviously, not all planes are flat, and the *z* axis at a given point cannot be considered square to a curved surface. That particular situation is accommodated by designating the *z* axis as perpendicular to a tangent of the curvature, often referred to as a normal vector, or "normal." These definitions and concepts can be at the heart of a new basis for understanding how forces and conditions can be described in a variety of situations.

A normal on a concave surface will diverge away from any other normal, while the same normal on the interior (convex surface) will converge and intersect with its neighbors. A sphere will tend to present a single normal in contact with a plane surface, while contact of a cylindrical body with the same surface will take the form of a line. The amount of contact is obviously determined by the amount of stress applied, the shape, and the elasticity of the bodies in contact.

This type of arrangement allows the surface of the contact to be defined within boundaries on the *x*, *y* plane and further defined by a *z* axis that extends into the contacting body (or from the center of mass when thought of as a direction of force). This simple attribution of locations is a starting point for calculations and measurements. Mathematicians and engineers can explain a great deal with reference to this model, which is known as an elastic half-space.[17] The model will accommodate some description of traction or friction with variations of a single force along the *x* axis. A combination of two or more forces would result in the calculation of a new vector that would represent the combination of force strengths and angles. The model could not be expected to replicate all situations, but it does provide a basis from which to examine many of the problems with forensic marks seen at crime scenes.

3.2.6 The Relevance of a Physical Basis

With the exception of direct transfers, almost all impression evidence is a product of contact between three-dimensional surfaces. This is a fact curiously at odds with the way in which most impression evidence is treated or presented. Representations of a three-dimensional product in a two-dimensional fashion logically suggest some loss of efficiency.

The basis suggested here is essentially a mathematical one that permits measurement. Measurement is necessary for assessing the nature of any contact.

Do parallel lines intersect? Bézout's theorem (1779) offered arguments to support that they do intersect at infinity but did not gain acceptance until the discovery of a method that allowed counting points at infinity. "As a result, Bézout's theorem, which turned out to be the main achievement of the theory of construction equations, was not properly proved until long after the theory had been abandoned."[3(p114)]

Science has become so widely accepted that, in many cases, the uninitiated defer to scientific sources over the immediate evidence of their own senses. When scientific predictions are seen to fall short of common expectations or hopes, the general public will often demonstrate dissatisfaction—ask any weather forecaster. There is no surprise that, in an ever more complex world, the basics of commonplace things, conditions, or abilities are routinely underestimated, while the underlying science is misinterpreted.

Acceptance of DNA analysis is actually one of the exceptions. The concept addressed a subject that satisfied all criticism:[18]

> In 1944, Canadian biologist Oswald Avery showed that DNA carried the genetic information within bacteria—but just how it did it, no one knew. In 1953, the American James Watson and Englishman Francis Crick figured out the answer. They saw that the DNA molecule was a double helix composed of two sugar phosphate strands linked by four repeating bases. (p. 19)

It is interesting to note that DNA was heralded as "DNA fingerprinting" in the early years of its use.

It is a shame when any evidence is abused or disused as a result of debate. Within the context of this book, the debate itself is moot. The purpose here is to initiate the reader to a more analytical examination of all the topics at hand.

The empirical method of reasoning used with DNA depends on the use of probabilities. There is the probability of uniqueness, which must also be accompanied by a probability of error. This is the same methodology used by the rest of the scientific world. It is used because, at least theoretically, it embraces the whole equation. Scientists have slowly come to embrace this method of reasoning in the past century, and it is now treated as a cornerstone of any new basis.

Edgeoscopy and ridgeoscopy ensure that no two details are precisely identical. The minute differences in the placement of pores and ridge units are significant when some other details can appear to possess some degree of similarity. This concept is a significant part of a basis that can be tied to the principles that no two things in nature are duplicated and no two randomly made markings are the same (with the codicil that they are not mechanically reproduced, such as with random damage to a manufacturing mold).

The growth of understanding relating to the fetal development of friction ridges does lend more weight to the simple concept that friction ridges are unquestionably individual. In the past, it might have been argued that some fingerprints have been found that are similar in a remarkable number of details. There is a flawed truth in the thought; it is possible to have second-level details such as ridge endings and bifurcations that roughly correspond in sequence over an area of pattern. It is also true that those same "characteristics" do not correspond at all in light of third-level detail comparison, such as of ridge units and pore structure.

In 2003, the landmark "Daubert challenges" questioned the validity of using deductive reasoning to reach a conclusion. It was in many respects inevitable that the question would arise. Scientists do not believe in absolutes. When faced with a single cause, they believe that there had to a huge error in the calculations. Their methods are precise, and for the most part, their calculations are generally proven to be sound; it is the finding of many small proofs that imparts some degree of reliability to even the most tragic findings.

Friction ridge practitioners frown on the use of probabilities to shore up an empirical statement. Practitioners cannot envision a meaningful set of probabilities that could be applied to friction ridge analysis. The courts tend to side with the scientific view.

There is, in all fairness, good reason to accept that inductive reasoning does offer a different approach to the same problem as deductive reasoning, and that if we could offer some form of error rate for friction ridge individuality, then both deductive and inductive methods could be applied to the cause.

The groundwork has been done, and as computer databases fill with criminal files and daily searches continue, the statistics are growing. The city of Sydney, Australia, has amassed approximately 100,000 digital criminal files since 1991. The city has a fairly low crime rate by comparison to many cosmopolitan areas. The figures, in an easily understood form, mean that each search of those files with a single crime scene impression has proven that either the marking exists in the database as attributed to an individual or information on the owner of that particular impression is not in the database.

This is not all that impressive until you realize that each file is composed of 10 digits. This means that the proven odds of finding two individuals with a single friction ridge impression in common must be greater than one in a million. This particular feat is being repeated daily. It is repeated in Canada, where the total criminal database exceeds 3.5 million individuals (36 million digits) and in the more than 100 countries worldwide with populations over 5 million and proportionately large criminal databases, none of which have reported a single duplicate friction ridge impression.

Each nationwide search by the RCMP (Royal Canadian Mounted Police), with or without an identification, adds another statistical straw to the pile of global results, none of which have implicated two sources for the same digit. These references are not the probabilities that result from formulas; they are just pure raw statistics, known as "descriptive statistics," that can be found in the search of any digital database.

Friction ridge impressions have a distinct advantage in the race to attain scientific acceptability in that the growth of databases reflects a continuously larger "sample population" from which to infer statistics. This may leave one wondering why the individualism of friction ridges has not been accepted. The answer lies in the very nature of a friction ridge: uniqueness.

This leaves a favored method of reasoning to be examined: abduction. This method, simplified, involves forming a best guess and then examining the guess to see if it can withstand scrutiny.

There is good reason to accept the actual process of individualizing evidential markings as a form of pattern recognition. The analogy is directly apparent with exhibits like footwear imprints and tire tracks, for which the pattern is what you expect to find. With friction ridge markings, one needs to look at a slightly more complex model.

The structure of the English language contains anomalies: "The soldier that deserts his desserts in the desert will receive his just desserts." This sentence is an example of just such a set of anomalies. English is notoriously difficult to learn as a second language, yet every child who learns to read English will have mastered the alphabet and the interpretation of the words to some degree in a relatively short time.

What is the mechanism that allows one to read? Imagine a san serif capitol letter *A* and a capitol letter *H*. These letters both consist of two long lines and one short line; despite this fact, one is readily discernible from the other provided that there is quality and sufficiency of detail (does this sound familiar?). The parallels between friction ridge marks and the written word extend beyond the way a letter is recognized.

Structural pattern recognition involves a hierarchy of pattern use based on pattern constructs that consist of simple units. There are only 26 letters used (in the Roman alphabet), yet a page of text from one book would never be mistaken for a page from another. Once a suitable level of proficiency has been attained, there is no common error rate for the act of reading. Reading has been mastered in different languages and forms around the world. It is a skill learned in school that is honed by daily usage; in fact, reading can be accomplished by computers. The translation of words to computations proves that reading, and by extension the translation of patterns, may one day become a machine process to at least some degree.

Simple applications of mathematics to impression evidence intuitively appear to represent a logical and achievable improvement to current practices. Such techniques come at a price in the form of time that must be invested with each step of the process, beginning with collection and measurement. The results offer the chance to improve the acceptance of those instances when they apply.

The need to accept the benefits of a system of identification is obvious in spite of flaws or shortcomings. The many topics considered in this chapter begin to describe the complicated but necessary new physical basis that can be applied to identification practices. Arguments for accepting or questioning identification practices tend to echo expressions of popular opinion regarding the inadequacies of an adversarial legal system.

Acceptance of less-than-adequate systems tends to foster resentment and mistrust. There is far too much waste invested in questioning that which need not be raised or allowed to become an issue. Both systems must change, and the renovation needs to originate from within to form a cohesive and unified new basis for identification and forensic testimony.

References

1. Verma, S. 2008. *The little book of maths, theorems, theories, and things*. New Holland Publishers, Sydney, Australia.
2. Bishop, M. A. 2002. *An introduction to chemistry*. Pearson Education, Lawrenceville, NY.
3. Stillwell, J. 2002. *Mathematics and its history*. Springer Science + Business Media, New York, 15.
4. Belashov, U., and A. Rosenberg (Eds.). 2007. *Philosophy of science contemporary readings*. Routledge, Boca Raton, FL.
5. Today in Science History. Dictionary of science quotations and scientist quotes. *Sidelights on Relativity*. 1920. http://www.todayinsci.com/E/Einstein_Albert/EinsteinAlbert-Quotations.htm (accessed June 13, 2010).
6. Bodziak, W. J. 1995. *Footwear impression evidence*. CRC Press, Boca Raton, FL.
7. Gregory, R. L. 1981. *Mind in science*. University of Cambridge Press, New York.
8. Galton, F. 1888. Co-relations and their measurement chiefly from anthropometric data, *Proceedings of the Royal Society*, London, England, 135–145.
9. Murray, R. C., and J. C. F. Tedrow. 1975. *Forensic geology, earth sciences and criminal investigation*. Rutgers University Press, Piscataway, NJ.
10. Gregory, R. L. 1981. *Mind in science*, University of Cambridge Press, NY. 34.
11. Sternberg, R. J. 1995. *In search of the human mind*. Harcourt Brace, Orlando, FL.
12. Gregory, R. L. *Eye and brain*. 1973. 2nd ed. McGraw Hill, NY.
13. Gregory, R. L. 1990. *Eye and brain, the psychology of seeing*. 4th ed. Princeton University Press, Princeton, NJ.
14. The Hartford Financial Services Group. 2009. *Aluminum wiring: Understanding the problem and its solution*. The Hartford, Hartford, CT. http://www.thehartford.com/corporate/losscontrol/ (accessed June 19, 2010).
15. Kirk, P. L. 1953. *Crime investigation: Physical evidence and the police laboratory*. Interscience, New York. http://en.wikipedia.org/wiki/Locard%27s_exchange_principle (accessed June 19, 2010).
16. Kies, D. 1995. The three appeals—Aristotle in the 21st century. http://papyr.com/hbp/appeals.htm (accessed June 19, 2010).
17. Johnson, K. L. 1985. *Contact mechanics*, Cambridge University Press, New York.
18. Daniels, P. 2007. *Body, the complete human*. National Geographic, Washington, DC.

4 The Ground We Walk On

4.1 Soil Composition

There is truth to the observation that "the facts should lead you," but that should not be used as an excuse to enter each scene with only a vague expectation of commonalities between scenes or outcomes of typical situations. Seeing a vegetable, we expect that it was grown in earth, but the plant it came from may have never been exposed to soil.

Soil is the natural matrix for plant life and a common substrate for impression evidence. Most soil contains a percentage of three basic substances: sand, clay, and loam. These substances can be easily separated in a direct representation of the ratio of components by a simple experiment.

A jar (such as a canning jar) filled approximately three-quarters full with a dug sample of surface soil and topped up to nearly full with tap water will, when shaken for about a minute, become cloudy and then begin to settle. After a short time (about 2–5 minutes), a discernible layer of precipitated material should be visible, and after about an hour there should be a second layer of precipitate forming a second band. The next day, the water should appear relatively clear, and the third and final layer of sediment can be measured. The three layers consist of sand at the bottom, fine clay in the middle, and loam at the top; the measurements provide the data for calculation of the approximate percentage of each layer (see Figure 4.1).

This experiment is relevant in at least three respects:

1. It would be good practice in a regional or even municipal setting to have some idea regarding the nature of soil types that could be expected to originate within the areas of interest. Soil composition (especially if known in advance) may offer some valuable investigative leads.
2. The description of soil composition could be part of a larger comparison of various soils at the same moisture content. In a specific case in which a "plastic" impression in soil (no matter the discipline) may be introduced in court, the quality of detail can be related to the composition of the soil.

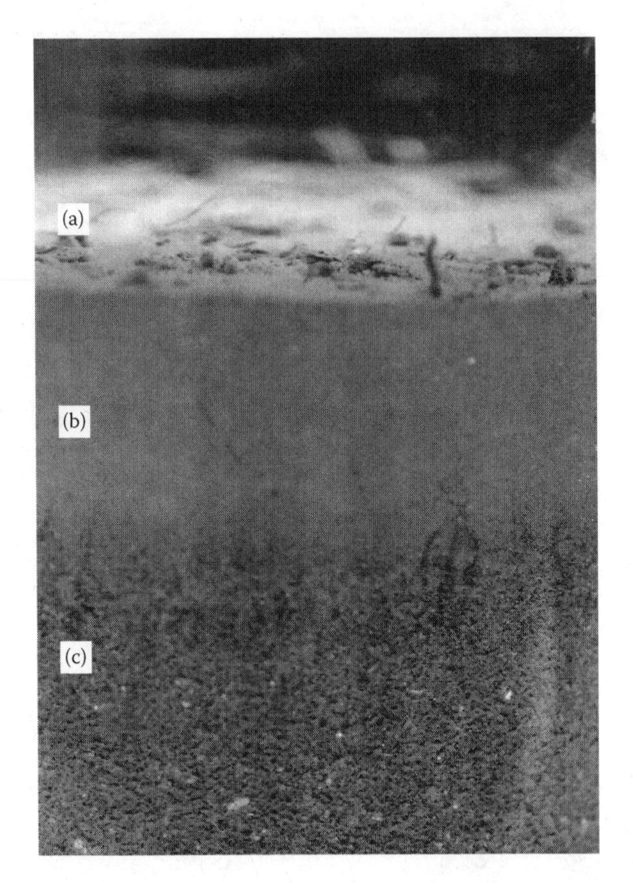

Figure 4.1 Soil composition by percentage. A soil sample in a canning jar has been allowed to settle for several days. This particular sample revealed a composition of less than 1% humus (a), 18% clay (b), and 81% sand (c). The clutter seen at the top of the image consists of debris resting on top of the soil.

3. Taken in close proximity to the questioned impression, the percentage composition is a descriptor that may assist in the evaluation of test impressions, perhaps explaining why one area of soil in a garden setting (for example) can be expected to yield impressions of either similar detail or depth compared to another area of similar impact.

If the question is a matter of confirming that a suspect impression could have been the product of jumping off a fence, then the soil should be tested. The testing may include a series of impacts by an individual of similar or greater weight wearing the same model and size of shoe. While the percentage composition of the soil could help, several samples should be tested, and soil samples for further, more exhaustive laboratory examination should be collected for submission. The suspect's shoes should be wrapped to preserve soil entrapped in the outsole, which can also be further analyzed.

There are methods of associating the depth of an impression to the weight of the source, but those calculations are based on the effects of mechanized equipment of great mass. There are no current means of reliably correlating the relative mass or weight of a vehicle or an individual, with or without load, to the associated depth of an impression.

4.2 The Structure of Soil

The size of particulate distributed in soil generally ranges between 0.06 and 0.002 mm (60 to 2 μm) for silt, less than 0.002 mm (2 μm) for clay, and more than 0.06 mm (60 μm) for sand.[1] The proportion of particulate size in a given sample will affect the size of voids between the constituents. Particulate size directly affects the water content of a specimen.

Water content is considered to be at a maximum value for agricultural purposes when the soil is fully wet and excess water has drained away. The volume of soil is affected by water content; contraction occurs during the drying process as the boundaries of aggregate are subjected to compressive force initiated by capillary action.

Clay offers some important properties of agricultural value that also have an impact on its usefulness in retaining details within an impression. Clay adds strength to soil in the form of coherence and structural stability (see Figure 4.2). With specimens of soil having concentrations in excess of 30–35% clay by mass weight, other constituents tend to take on the properties of the clay content.[2]

Farmland with clay soils tends to become waterlogged with excess rainfall, stay wet longer than other soils, and form large clods that are routinely carried from the source to another location. The clods can provide soil samples for further analysis by an accredited lab or form a trail of miniature casts showing the path of a suspect as the matrix dries and releases from the outsole.

Practitioner's tip: Working a scene can be a harried event and evidence in the form of small clods of earth may be easily destroyed or forgotten. Your attention will usually be required elsewhere. The procedure should be to photograph the pieces in situ and then collect them in one or more sealed and labeled containers (to preserve the moisture content) to further secure them from damage in transit and to continue your examinations under more controlled conditions. Remember that they might be the only footwear or tire impressions that you find.

A shoe or tire in contact with soil affects the particles beneath it in a predictable fashion. It is not practical in forensic matters to attempt to map the effect of forces on particles since the shapes of the particulate are nonuniform. What can easily be observed and possibly measured is the cumulative effect of an amount of pressure minus the displacement effect on the surrounding mass (see Figure 4.3).

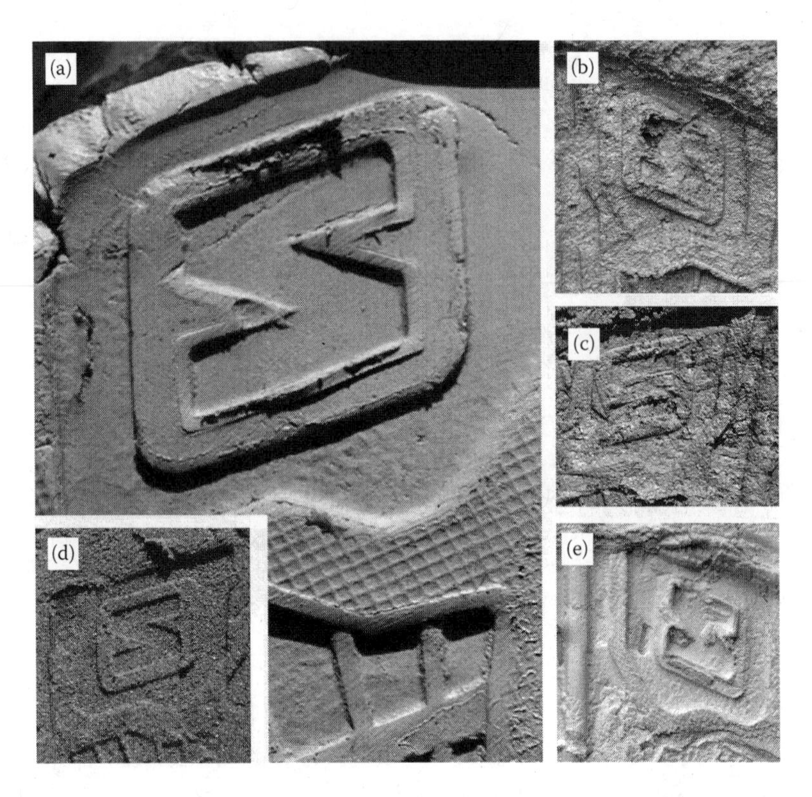

Figure 4.2 Comparison of footwear outsole impressions. (a) A footwear impression in clay; (b) and (c) images taken with the same soil as seen in Figure 4.1; (d) impression in damp sand; and (e) impression taken in talc. Impressions in (a), (d), and (c) were all made with the left shoe, whereas those in (b) and (e) were made with the right shoe of the same pair. (a) Demonstrates that clay offers a superior substrate for recording impressions.

Considering the loads from multiple sources and that the exact conditions cannot be easily calculated, one might easily be discouraged from thinking that a study of soils has any practical value. The fact is, however, that the discrete particles in soil do tend to compact into a relatively continuous medium. This means that even coarser materials such as sand can retain useful details, particularly with the presence of some moisture to provide cohesion between the particles.

Compaction of a soil during contact, particularly with a coarse specimen, will force both linear and rotational motion of particles. The particles will tend to congregate as the voids between them are compacted as well. This results in the presentation of a much smoother and tighter arrangement of particles at the surface.

In theory, this topic may one day be tackled by computer analysis. The parameters are measurable, and one would expect that, with the development

Figure 4.3 Displacement of modeling clay. A cake of modeling clay was impressed to create a waveform pattern of parallel squared ridges (as viewed in cross section). The clay was cut in two, and a small jar was impressed into the center of the clay on the left with a force of approximately 69 KPa (about 10 psi).

of software and sampling methods, predictions could be made that would produce soil analysis according to characteristics, including evaluations of compressibility.

4.3 Influences that Act on Soils

The conditions under which a soil is formed can greatly influence the disposition of a soil. A sedimentary soil that is subjected to loads beyond those imposed during settling will compact in response to the amount and duration of loading. Sediments that have settled slowly over a large area without the incursion of other disturbances may be expected to exhibit symmetry in structure and mechanical characteristics.[3]

The geography of a region and consistencies between specimens provide mere starting points for understanding the nature of soil at a crime scene. In addition to natural attributes, there are the activities of gardeners, the source of foreign topsoil in suburban areas, and even city planning initiatives to be considered. Where open areas are affected, soil erosion from one area can well affect another. These considerations must be taken into account in a thorough evaluation of crime scenes.

The amount of information available in a particular soil can usually be expected to correspond to the amount of moisture content in the sample. Wet soil is considered to have elastic and plastic properties, complete with identifiable shear rates, while the same soil without water will exhibit significantly less cohesion. Understandably, the ability of a soil specimen to retain recorded detail is easily degraded by the drying effects of temperature or wind, with at least some examples.

Soil exists in a state of constant stress, in which gravity is ever present and the flux of moisture content regularly alters the forces of cohesion within a local sample. The stresses are time dependent and susceptible to change, not only from pressure, temperature, and water availability but also from chemical and biological activity.

Not all influences have a negative impact. In a concrete jungle or a highly compacted stone driveway, one does not usually expect to find a viable example of footwear impression in soil; yet, this is made possible by the activities of ants. There are two fairly common species (small black ants and pavement ants) that tend to provide the means of obtaining impression evidence in what would otherwise be considered a hostile environment (see Figure 4.4).

The images in Figure 4.4 depict the activities of pavement ants *(Tetramorium caespitum)*. The species was identified based on observations of their activities, size of workers (about 3 mm), brown color, striations observed on the head, the presence of two nodes between the thorax and the abdomen (seen on at least one specimen), a 12-segmented antennae, and that the antennae ended in a three-segment club. These attributes compared favorably to published descriptions of pavement ants.[4]

Pavement ants were introduced to North America from Europe and are now widespread throughout most of the continent. The nesting and burrowing activities of pavement ants create a relatively fine substrate in places where none existed before their contribution. They are not the only creature that may be affecting the topography of a crime scene in this fashion.

Another potentially significant influence on the structure of the soil and surface conditions may arise from the action of worms.[5] The thought that worm castings may have an effect at scenes of crime is not supported here by findings but only presented as theory. If the account of the activities of British worms has transferred to local species that migrated from Britain, then their efforts should be observable, and with a cold case, they may have a considerable impact on findings.

Darwin's[5] accounts detailing the accumulation of layers of castings that could potentially bury objects over the period of a few years may well pique the curiosity of investigators. This thought inspired me to contact the author John W. Reynolds.[6] In conversation, he indicated that *Lumbricus terretris* and *Aporrectodea longa* would be the most active of our local species and therefore a good candidate for study in the future.

Equipped with some knowledge of what could happen, one is much more likely to conduct a thorough examination. This also applies to protection of

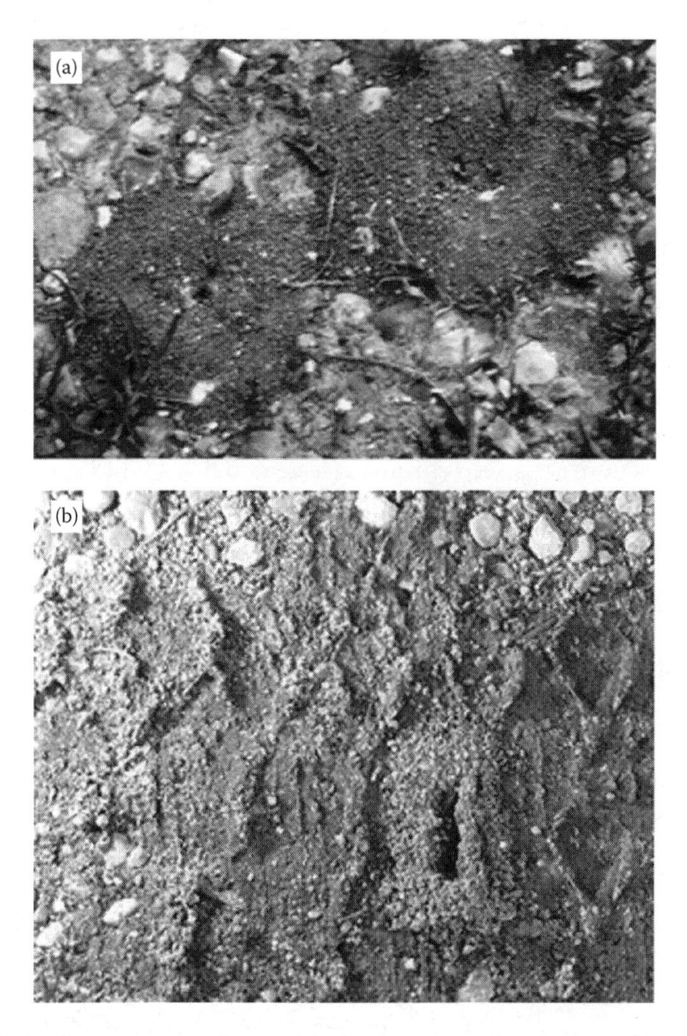

Figure 4.4 The activities of ants. (a) The two mounds of extracted particulate are the by-product of burrowing activities. The aggregate is carefully placed with an interesting degree of symmetry in a circular fashion in this undisturbed portion of stone driveway. (b) In another area of the same driveway, a light rain is responsible for the darker tone of the older surface materials. Notice that the more recent activity is lighter, and the dispersion pattern has been affected by the pattern of the tire track that had been made approximately 2 hours before this image was taken.

a scene where local weather patterns, topography, or susceptibility to damage may dictate both the order and the priority of searching methods. Just as a laboratory should have protocols for testing the validity of a process, so should field-workers seek to establish and update accepted and documented protocols for the treatment of local scenes under local conditions.

4.4 A Curious Destination for Minerals and Dyes

There are various uses for many natural materials, including minerals and dyes. The substance referred to as magnetite (Fe_3O_4; a form of iron oxide), also known as ferrosoferric oxide, was determined to be the active and predominant substance in a black magnetic fingerprint powder. This substance ironically appears in a list of possible ingredients used in the tattoo industry.

Many people that pay substantial amounts of money for the honor of having their skin decorated are unaware of the potential list of substances to which they may have been exposed. The industry is unregulated, and there are no ingredient lists available for reference. Some of the materials may include plant dyes, rust (iron II oxide), carbon, copper, cobalt aluminum oxides, aluminum salts, lead chromate, lead carbonate, admixtures of potassium ferrocyanide, or ferric ferrocyanide.[7]

These and other substances do not sound enticing and may exhibit varying degrees of permanence in use. The question of permanence with tattoos on friction skin will obviously vary with the substances used to "fill in" the tattoo. One individual displayed a tattoo of paper currency on the center of his palm, received about 20 years earlier. He claimed that it was incomplete without the matching tattoo of an eyeball that he had neglected to have placed on the other palm, so that whenever cuffed, he would be able to keep an eye on his money.

References

1. Bardet, J. P. 1983. *Application of plasticity theory to soil behavior: A new sand model.* Technical report CaltechEERL: 1983. SML-83-01. California Institute of Technology. http://caltecheerl.library.caltech.edu/321/ (accessed June 4, 2010).
2. Newman, A. C. D. 1984. The significance of clays in agriculture and soils. *Philosophical Transactions of the Royal Society of London A,* 311, 375–389.
3. Muirwood, D. 1998. Life cycles of granular materials. *Philosophical Transactions of the Royal Society of London A,* 356, 2453–2470.
4. Harvard University, University Operations Services. 2006. *Pavement ants (identification, biology, and control)* http://www.maine.gov/agriculture/pesticides/gotpests/documents/pavement-ants_harvard.pdf (accessed June 4, 2010).
5. Darwin, C. 1897. *The formation of vegetable mould, through the action of worms, with observations on their habits.* Appleton, NY.
6. Reynolds, J. W. 1977. *The earthworms (Lumbricidae and Sparganophilidae) of Ontario.* Royal Ontario Museum, Life Sciences, Toronto.
7. Halmenstine, A. M. 2010. *Tattoo ink chemistry.* About.com Guide, About.com, a part of the New York Times Company. http://chemistry.about.com/library/weekly/aa121602a.htm.

Measurement

5

EUGENE LISCIO
DAVID S. PIERCE

5.1 The Relevance of Measurement

5.1.1 Introduction

Crime scene and evidence measurement have improved in recent years and can be expected to improve further. The regular use of any type of scale to measure impression evidence has only become standard practice at crime scenes in the past few decades. It is a rare sight these days to see a duty book, coin, or pen used as a substitute for a scale. This change is a good thing.

Measurement, developed by many societies, can be traced to ancient examples. The use of some form of measurement allowed problem solving that includes the most basic operations, such as division. The importance of these activities could be seen to influence the success of a society that, by employing rationing or tracking herd populations, for instance, would improve its collective chances of long-term survival.

5.1.2 Measurement and Counting

Measurement of anything is generally based on certain concepts, such as counting, taught in grade school. Mention of counting numbers is commonly associated with base 10. Counting is, however, a language in which the base 10 decimal system is but one dialect. Reference to a computer data size of 1,024 appears to be a randomly chosen or mystical figure until one realizes that it is derived using base 2, denoting 2^{10} (= 1,024). "Similarly, mega-, giga-, and tera- are applied to a base 2 and denote 2^{20} (= 1,048,576), 2^{30} (= 1,073,741,824), and 10^{40} (= 1,099,511,627,776) respectively."[1(p163)]

Scientific notation (a method of making large numbers easier to handle) is expressed in powers of 10. Engineering notation differs in that the powers are expressed in multiples of three, so that 10^6 (M, mega), 10^9 (G, giga), and 10^{12} (T, tera) are used, but 10^7 does not appear in engineering notation.

The application of numbers as a method of measurement is a commonplace tool in forensics. Measurement is descriptive, allowing some estimation of size, shape, or other discriminatory information. The legal system then makes judgments based on such information, which can be regarded as corroborative or of assistance in the interpretation of evidence.

Relevance is routinely weighted by the courts, thus creating a need to prove adequacy, accuracy, and precision for each case. Methods will continue to change with developments in technology as a natural course of affairs. The use of measurement is an attempt to create an exact, accurate, and precise representation of evidence, situations, or events.

Geometry is a form of counting in which points, lines, and angles become measurable quantities. In use, the definition of lines, circles, and angles come alive as shapes with application of Cartesian coordinates. This language is elegant in its ability to define complicated matters without ambiguity.

5.1.3 Thresholds and Proportion

The world we live in exists largely beyond our ability to sense it. This refers to the thresholds of perception that confound our senses. Presented with an image of a bright speck on a dark background, we would have no means of determining whether it was an image of a star, some microscopic sample, or something between those extremes.

Many thresholds can, do, and should apply to forensics. The approximate capability of a person with 20/20 vision allows resolution of image detail down to about 0.076 mm. Objects or images outside our range of vision require an application of technology that must be justified when tested in a court of law.

Our detection abilities are further complicated by the wide span of evidence dimensions and the limitations of the equipment used to enhance it. There are four scalar divisions used for evidence from visible, macroscopic, microscopic, and molecular ranges. Measurements that provide adequate reliability and accuracy within one range are often not practical for another application.

5.1.4 Measuring Equipment

Reticules, used to measure features under a microscope, are seldom useful without the tool they are designed to complement. Similarly, an ABFO (American Board of Forensic Odontologists)-style scale works well on medium-sized features and tends to reveal both distinct and minor flaws inherent in the scale with greater degrees of enlargement in macro photography.

The capacity of a computer to record with precision is on the order of 0.42% relative error at 300 dpi (dots per inch) or 0.32% relative error at 800 dpi when measuring a feature 10 mm long (less than ½ inch). These levels of performance require accuracy in the scale used to calibrate the computer. The results (in pixels per inch or pixels per 25.4 mm) will reflect and amplify any imperfections in the source of calibration.

No single tool can be expected to fulfill all requirements. Well into the nineteenth century, scientists were renowned for their ability to devise their own equipment. Glassblowing skills were historically essential abilities, particularly for chemists.

Tools that are part of experimentation need to prove their worth. A contour gauge, used to replicate a cross-section of topography, is a contact instrument. As such, it would not be recommended for forensic use except as a last resort. The noninvasive use of a point cloud serves much the same purpose only with far greater accuracy and precision. A typical use could include imaging a particular section of an outsole or tire tread to reveal the cross-sectional shape. In the matter of a physical match for an otherwise difficult to illustrate set of objects, even a contour gauge could be of some use during an analysis or comparison.

Common use of noncontact methods, such as the use of laser scans (point cloud) or photogrammetry, can be expected to supplement or replace contact-based methods, perhaps one day even usurping the roll of casts.

A number of devices were employed or experimented with during the writing of this text. Many of the tools used are not commonly found in a lab setting, a forensic office, or even outside a private collection (see Figure 5.1).

Tools that were omitted from this image include a film thickness gauge, reticules, and a set of sheet metal and wire gauges. Items included that would not likely be recognized or readily accepted by most practitioners include the following:

1. Three calipers, two of which are accurate to ±0.01 mm
2. Spreading bars 609.6 mm (about 2 ft) long for creating films with liquids such as paint or ink in predetermined thicknesses
3. Three 25 × 50 mm cut metal scales illuminated by a flashlight
4. Pocket weighing scale accurate to 0.1 g
5. Test strips for pH and various flat scales and rules, including metal, plastic, and paper versions
6. Retractable pocket magnifier at about ×4 enlargement
7. Contour gauge (currently displaying a cross section of the outsole of a boot)
8. ABFO forensic scale
9. Scientific calculators

This attention to the variety of tools used in some forms of research is meant to illustrate the number of devices that can be applied in forensic examinations. Contour gauges, film spreaders, and pipettes (not shown) or calipers are borrowed technology. Advancements that produce useful

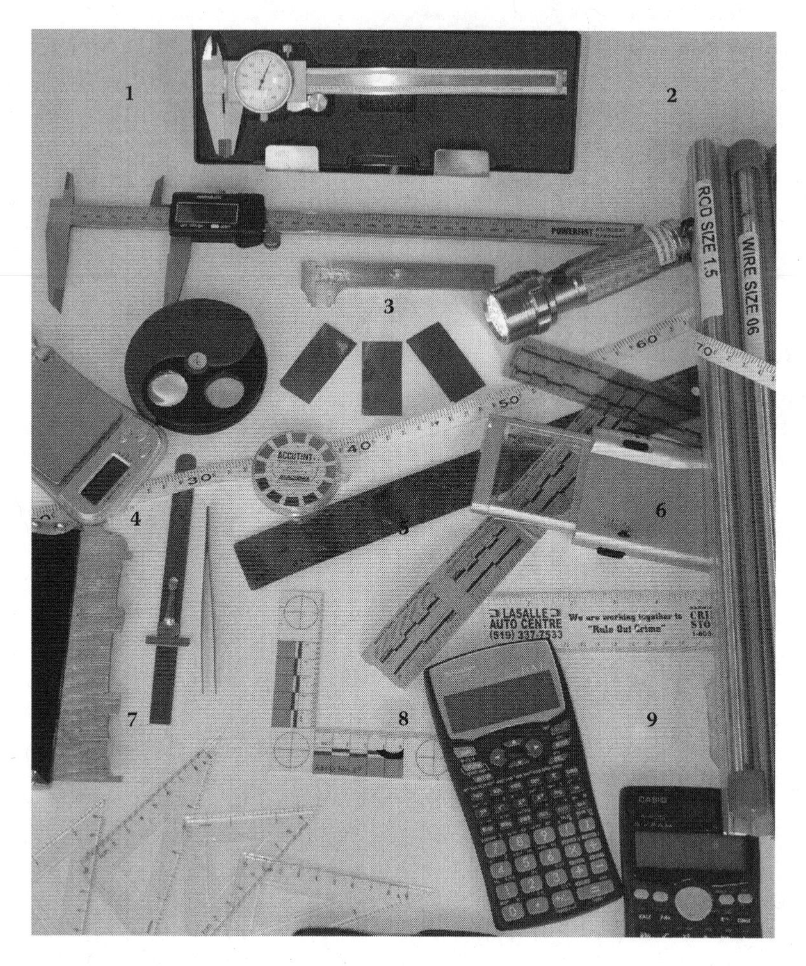

Figure 5.1 Measurement tools.

terminology, practices, or equipment can and should be adapted for forensic use.

5.1.5 Precision and Accuracy of Measurement

Simply put, accuracy depends on how well made the scale is (exact separation from one unit to another), while precision is determined solely by the size of the increments and not their quality. In the case of scales currently in use to photograph evidence, the error of the scale is transferred to any subsequent image examined in, saved on, or produced by software.

The software typically used to calibrate the image is capable of measuring in pixels. When an image is downloaded and calibrated (into pixels), the calibration is based on the smallest fractional increments of the scale. This means that if you choose to create or save an image at 100 pixels per unit of measurement (i.e., per inch) and another at 1,000 pixels (per inch), the second image will naturally have significantly greater precision (here is the important part) but only precision. If the unit is calibrated with some inaccuracy, that inaccuracy will be transferred or amplified despite the presence of precision.

Error, transferred with measurements made with a defective scale, is generally not a mistake but rather a level of uncertainty. When the absolute value of a measured quantity is known, it is possible to determine a reasonably exact measurement of the amount of error. With most forensic matters, the absolute value of an item being measured is not known, and any method that can minimize error, especially with a source that can be attributed to a manufacturing flaw, is a welcome innovation.

Practitioner's tip: You can easily test the accuracy of a particular scale by taking a photograph at 1:1 or closer. Just load the image into a suitable editing application and make sure that the image is horizontally level and at a high degree of magnification (not more than about 16 mm in view). Then, create a box with the appropriate selection tool so that the length starts and ends at the exact center of the marks indicating a 1-mm increment. Dragging the box you made with a "move" tool will allow you to evaluate the accuracy of other 1-mm markings. If your needs for a particular image will not require this level of concern, you may yet be well advised to evaluate your measurement instruments.

In perspective, the overall measurement of a footwear impression or tire impression using a scale that measures in millimeters is appropriate, and the error rates for those overall measurements are not significant. The need for greater accuracy and precision with those scales arises with finer measurements of features within the impressions, such as scratches, gouges, or inclusions, like stone holds.

5.1.6 Introducing a New Scale

It is a common goal to strive toward achieving the best possible result with the equipment and conditions at hand. Customary practices that yield serviceable results may cause one to regard suggestions of change with great suspicion.

The relative error of any measurement is a measure of precision in which the greatest possible error is divided by the measured value, and the result is expressed as a percentage of relative error.[2] The greatest possible error is accepted as one-half of the smallest increment. During the research for this book, a metal scale with etched markings was manufactured, with its smallest increments 0.25 mm. Table 5.1 shows the relative error rate for a typical forensic scale (mm) and for the metal 0.25-mm macroscopic scale.

Table 5.1 Error Rate of Forensic Scales (Comparative Precision) (%)

Length of Feature (mm)	Printed Plastic/Paper Scale Increments (1.0 mm)	Cut Metal Scale Increments (0.25 mm)
1	50	12.5
5	10	2.5
10	5	1.25
15	3.3	0.83
20	2.5	0.63

Table 5.1 further illustrates that the macroscopic cut scale performs best when used to examine small features. The gap between the precision of these pieces of equipment diminishes with increases to the size of the feature examined. The concern here is a matter of deciding what size product is required for a specific task. Printed scales work just fine for average features that measure in the range of a few centimeters or more.

The tabled comparison of relative error between the two scales illustrates predictions of the precision of any two scales of similar increments no matter what other attributes or materials are involved. The image (Figure 5.2) of different scales illustrates specific accuracy issues that are common only to these portions of these two scales (there are slight differences in the distance between the millimeter marks on the printed scale).

The metal scales can be chosen in a range of thicknesses that rivals the use of a paper scale when focal plane issues arise. A typical printed scale measures about 1.02 mm in thickness for plastic, to 0.08 mm for an adhesive paper scale removed from its backing material. The use of a thicker scale may require that steps be undertaken to ensure that the scale and the exhibit are on the same focal plane. Cut metal scales (as shown in Figure 5.2 and again in Figure 5.3) can be manufactured in different thicknesses (the ones shown measured 0.13, 0.25, and 0.45 mm as measured with a digital caliper accurate to ±0.01 mm).

Cut scales do have drawbacks; for one thing, they are opaque. One difference between the cut scales and printed metal scales is that the cuts penetrate the thickness of the material, thereby allowing a more precise alignment of the markings to fine features. A typical and inexpensive stainless steel printed ruler is available in a thickness of about 0.45 mm but is subject to focusing issues and wear of the printed marks over time and use.

Clear plastic scales as found in math sets offer a potentially useful attribute in that they are often printed in reverse on the underside of the material. This means that the scale may be placed on the same focal plane as the uppermost portion of the subject, thus alleviating the thickness issues. Unfortunately, this also exposes the printed image to whatever surface irregularities are present.

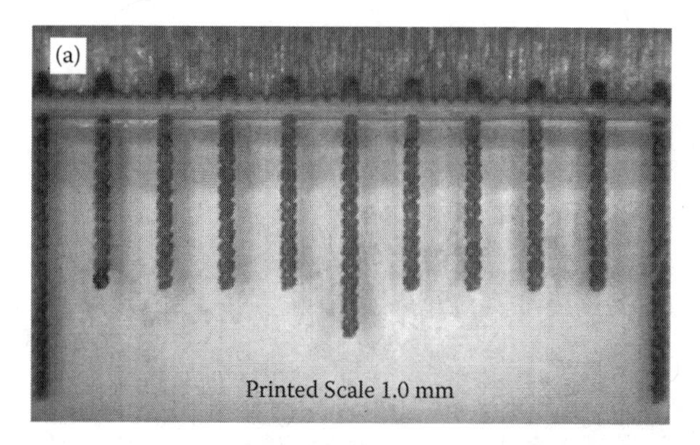

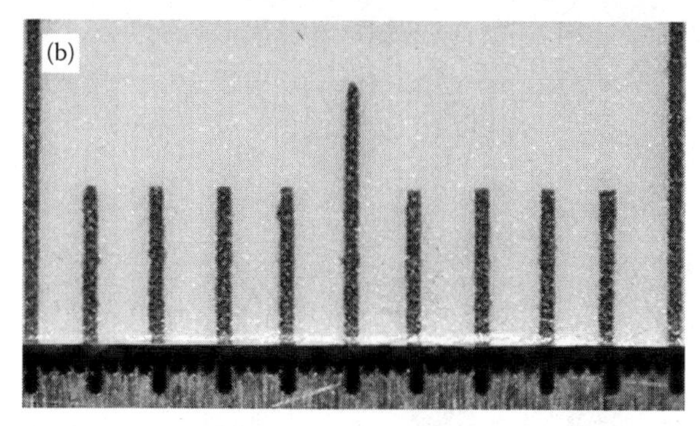

Figure 5.2 Comparison of three scales. (a) The markings of a clear plastic math set ruler closely match the 1-mm marks of the cut scale (top edge of image). (b) A simultaneous comparison of the markings printed on a plastic forensic scale compared to the cut scale (bottom edge of image). Note that the cut scales divide 1-mm increments in four.

Problems with the use of these scales include the need to keep the piece clean and lubricated since even stainless steel (of a lower grade) will rust if not protected. The care to avoid damage extends to ensuring that the scale is not placed under excessive stress; metal can become deformed (bent or dented). Finally, the finely cut edge must be treated with respect; it poses a danger in that it is a sharp object and can cut through skin easily (see Figure 5.3).

It should be noted that the single measurement of any object with a scale would result in a two-dimensional (2D) representation of a particular length. Care must be taken to ensure that comparative measurements address this issue by using common points of reference.

Questions about whether to focus on the exhibit or the scale can become moot when both are in focus. The exception to that statement arises in consideration of an exhibit that does not permit ease of focus, such as a dark marking on a dark substrate, for which the scale may be the only practical

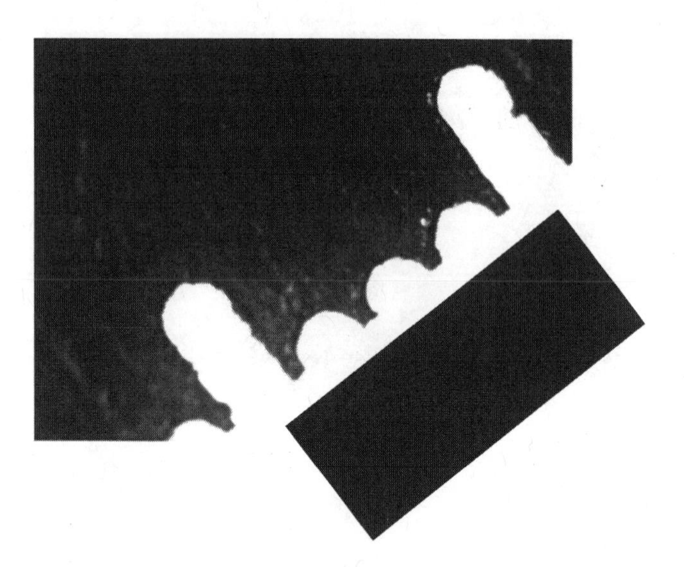

Figure 5.3 Teeth of a cut scale. In this enlarged figure, the black rectangle was computer drawn and is 1 mm long. The image demonstrates the structure of the cut scale at a reasonably high degree of magnification and points out that the teeth require cleaning to control a buildup of debris, such as fibers and small flecks of other foreign materials.

surface for focusing. In fact, where the scale has been placed as diligently as it should be, there should be no difference in focus.

Cut metal scales should likely be used both individually when focus and precision are critical and as an adjunct or attachment to a printed scale if the image will be used for calibration purposes. There is little difference about the method of use for a cut scale except that current models lack circular features that assist in an approximation of parallel orientation of the film plane to the subject plane (thus controlling parallax distortion).

In the instance of a deep imprint as found with many footwear or tire tracks, close detail requires the scale to be positioned on the same focal plane as the intended subject. This means that the automatic controls of modern cameras may need to focus on the upper regions of the impression even though the relevant evidence is recorded only in the deepest portions of the impression that correspond to the outer surface of tread patterns. This focus management issue is best solved by focusing on a scale that has been placed at a depth to allow appropriate focus.

This is another technology that may one day be supplanted or replaced by an improved method. It is entirely possible that sophisticated imaging software (with automatic measurement) could be developed that would potentially eliminate the need for any form of scale. Imaging software is dependent on the pixel, so in turn software could be similarly made redundant by cheaper, more efficient laser scanning.

5.2 Accuracy and Language Issues

5.2.1 Introduction

The identification of friction ridge impressions provides a few good examples of measurements that can be flawed. Consider a set of known markings from the same source. Some have been recorded with ink; others are digital records. A scene-of-crime marking is a partial imprint that has just enough detail to allow a count of the intervening ridges between one feature and another. While the ridge count (number of ridges intervening between one feature and another) is consistent between all of the known originals, the scene-of-crime marking, although otherwise consistent, is off by just one ridge.

An experienced practitioner will immediately recognize and resolve the issue without trouble. The scene-of-crime marking should be checked for other signs that it is tonally reversed. The number of ridges counted is often off by just one ridge. The error is a bias of measurement, akin to repeatedly measuring a person's height only to realize that one measurement was made with the suspect's shoes off while the others were taken with shoes on.

This concern for accuracy is reflected in the principles of identification that stipulate that there can be no unexplainable differences between impressions for the purpose of individualization. Misunderstanding the significance of unexplainable differences can lead to doubts for the need for principles of identification. Imagine that the possible circumstances under which comparison of two features of any friction ridge imprint could be different by a single ridge. These circumstances include tonal reversal, practitioner error, distortion, flexibility of skin, addition of a digital artifact during scanning, or the possibility that the two markings originated from different sources.

One cannot make useful measurements in the presence of doubt. The value of explained differences is that they remove that doubt. This also applies to such influences as interference and interventions (in which some factor in a situation causes artifacts to be created) or dimensional stability.

Clues such as the occupation of a suspect or knowledge of activities can allow one to interpret such anomalies. These issues are largely systemic and can easily be resolved if the problem can be identified. Issues such as the misrepresentation of height can be statistical problems that in a closed system (i.e., a particular location) may be reduced to a systemic solution but take, for instance, a set of measurements from different locations.

Different methods of recording the same class of data may cause flaws that result in a range of measurements or, worse yet, conflicting data. Science and mathematics tell us from an early age that counting is more accurate than measuring, and that no measurement can be considered error free, no matter how carefully it is made. Forensic accounts of evidence include some forms of counted measurement, such as in the counting of friction ridges or the pitch sequence of a tire.

5.2.2 Measuring Waves

There is room to improve the descriptive accuracy of measurements. In fact, there are applications of measuring techniques that could be adapted from other sciences to suit specialized forensic purposes. One example is suggested here regarding the description of a wavelike pattern that could as easily be found in a tool mark, footwear impression, or the transferred details of some foreign patterning (Figure 5.4).

Electromagnetic waves are described first as waveforms. This refers to the type of feature commonly seen on the screen of an oscilloscope. While those waveforms are meant to define the variation of a quantity with respect to time, a static pattern is simply concerned with the relative amplitude and frequency of generally consistent phases. When the signal, or shape, repeats at regular intervals, it can be said to be periodic or continuous. A unique and nonrepeating example, by contrast, can be described as aperiodic or transient.[3] Waveforms are traditionally measured as a single phase unit. The first measurement is taken from a theoretical rest line that runs parallel to and equidistant from the line suggested by connecting the uppermost points of crests and, separately, the lowest points of wave troughs. The measurement of wavelengths λ can be taken from any corresponding nodal point along a

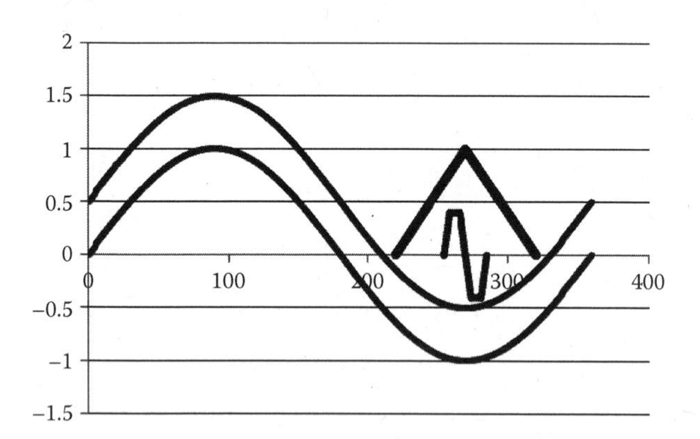

Figure 5.4 Graphing of waveform patterns. Two complete sine waves: the top half of an angular wave pattern, and one complete waveform that appears truncated. The sine waves were plotted based on calculations made using a handheld scientific calculator and manually entered into a spreadsheet. A graphing calculator would have provided a better product much quicker, but this illustration was created using commonly available equipment. Spreadsheets that calculate in radians will require a simple adaptation to plot from an equation.

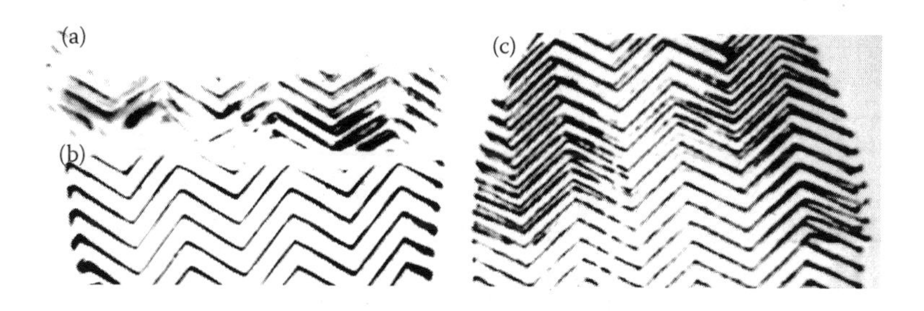

Figure 5.5 Waveform distortions. The susceptibility of a new shoe to deformation stemming entirely from pressure application. The same shoe made each of the impressions, before and during bending. There is an observable difference in width between the impressions of (a) bent at contact, (b) made in the restive state and the two overlapped impressions, and (c) a restive marking overlapped by the same outsole while bent.

wave, but the three intersections along the rest line (start point, half-wave point, and full-wavelength point) are likely the best choice for accuracy (see Figure 5.5).

A practitioner who accepts some unified definition of waveform structure may further appreciate some accompanying definition of distortion since that also is quantifiable. The definition offered in the study of electronic circuits identifies a difference for which the output is not identical to the input. This simple adjunct to our knowledge of the potential for some materials to behave in an elastic manner gives reason to think about a unified method of describing the differences between *input* and *output*.[3]

Distortion associated with the application of some kinetic force will most often show directionality, as seen with scufflike or parallel striated markings. The direction of force should permit the examiner to determine the presence of distortion in terms of an *amplitude distortion* or *frequency distortion* as the dominant response to the force applied. Remember that although a shoe, for instance, may show a more limited capacity for distorting across the width of the outsole, the substrate of a pairing may demonstrate no such limitation.

Examinations in which distortion is to be quantified may be guided by directional clues and may begin with measurements of amplitude and wavelength for each of the features seen on the crime scene impression, the suspect tool (or shoe), and a series of test impressions. Further associations, because the outsole may be compressible, could include measurement of the compression of specific areas of the outsole that exhibit details proportionate to or consistent with the morphology of the foot. Likely stresses, such as the bending of an outsole when standing on tiptoes or the curvature of edge characteristics in a slide, can further be the root cause of distortion.

While these determinations are merely consistencies, they may assist in an analysis. Inconsistency that has not been quantified amounts to an unacceptable variable. The only prediction that can be made is that looking for minor variances to prove consistency is a hallmark of a job well done.

The presence of a measured discrepancy will need to be repeatedly confirmed. The measured difference between input (the condition of the outsole at rest) and output (the test condition under stress) should offer some insight regarding the value of a hypothesis that accounts for the apparent distortion. This technique applies to any reliably measured features of any pattern.

The advantages of the suggested terminology are believed to be accuracy in the use of unified terminology and the provision of a method by which to quantify an amount of visible distortion. Even distortion that obscures some significant detail may be worth study. When distortion is found within the boundaries of an impression, a practitioner is well advised to be able to explain or quantify the area of distortion with a similar amount of authority as applied to the areas of congruence.

5.3 Photogrammetry

5.3.1 Applications

The digital camera is by far the most used instrument to document and record crime scenes for law agencies all over the world. These are relatively inexpensive, easy to use, and durable; operate in varying environmental conditions; and can provide high-resolution images that can be stored indefinitely with no loss of quality. In addition, digital still and video cameras have found their way into cell phones, laptop computers, transport vehicles, security systems, and even police vehicles, which are being retrofitted with dash-mounted video cameras. As a result, there has been an exponential growth in the amount of data captured digitally each day, and there are more crime and accident occurrences caught on camera than ever before.

However, it is surprising to note that most people are only remotely aware that through photogrammetry they can use a digital camera as a sensitive measurement and surveying instrument. Also, photogrammetry allows an investigator to use photographs taken by a third party to determine measurements for objects in a scene. There is no other technology that allows a person to go back in time and recover measurement data from a crime scene photograph.

Photogrammetry has some of its earliest roots in the use of perspective geometry for paintings and art during the time of Leonardo da Vinci around the year 1480. During this time, da Vinci stated the following:

> Perspective is nothing else than the seeing of an object behind a sheet of glass, smooth and quite transparent, on the surface of which all the things may be marked that are behind this glass. All things transmit their images to the eye

> by pyramidal lines, and these pyramids are cut by the said glass. The nearer to the eye these are intersected, the smaller the image of their cause will appear.[4]

This early description of how light reaches the eye by "pyramidal lines" is accurate and is often shown in simplified drawings of how light enters a lens and is projected on an image sensor.

The principles of perspective and projective geometry form the basis from which photogrammetric theory is developed. However, most of the growth of photogrammetric techniques, equipment, and methods has been closely tied with the use of aircraft for surveillance and reconnaissance, especially between World War I and World War II. During World War II, virtually all available photogrammetric mapping facilities, civil and military, in all the belligerent nations were assigned the formidable task of producing a vast amount of maps and charts to meet innumerable exigencies of war.[5]

The use of photogrammetry in crime scene reconstruction was mentioned by the American Society of Photogrammetry in the third edition of the *Manual of Photogrammetry*[5] as early as the 1960s. Since then, photogrammetry has branched out into the digital realm and covers many different techniques for measuring objects that range in size from micrometers to many kilometers. With such versatility in range, photogrammetry has found uses in archaeology, architecture, medicine, robot vision, biometrics, and even underwater applications. Typically, crime scenes fall in the category of "close-range photogrammetry," which encompasses imaging and measurement in the range of several centimeters up to hundreds of meters.

The term *photogrammetry* can be broken down into its components of *photos* as in light, *gramma* as in diagram, and *metry*, which is the science of measuring. So, what we have is the science of gathering measurements from photographs. There are more specific definitions of the term that include remote sensing and other technologies and concepts, but at the heart of photogrammetry is the notion of being able to obtain geometric information about objects in a photograph. The photograph may not necessarily be a digital image but may include paper print, film (negative), stills from analog video, x-ray, and any other media used to store images.

When we take a photograph, we collect the photons (rays of light) that are present in the view of the camera. These photons either emanate from a light source or reflect off objects in our three-dimensional (3D) world. As we collect these light rays through the camera lens, they converge to a point known as the focal point; subsequently, the light rays are projected from our 3D world onto a flat 2D plane, which is usually the imaging sensor on a digital camera. Photogrammetry can be thought of as the process by which we

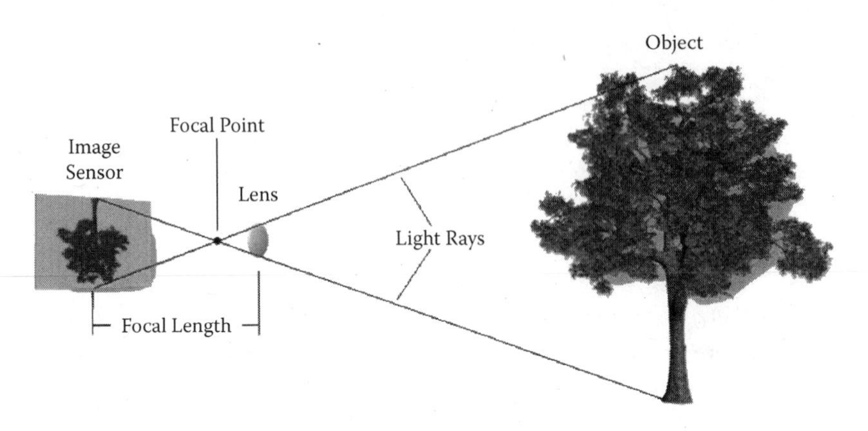

Figure 5.6 Simplified camera setup.

do the reverse. By knowing some information about the camera parameters, its position and orientation, we can work backward to calculate the origins of the light and hence the 3D positions of objects in space (Figure 5.6).

There are normally two circumstances that help define the methods, techniques, and possible information that can be extracted in any photogrammetric project. The first circumstance is when a person is going to be present at a scene and will be taking the photographs with a "known" (calibrated) camera. The second situation is when photographs have been taken by a third party, and the original camera is unavailable or "unknown" (noncalibrated). This second circumstance includes images on any media, such as photographic prints, film negatives, and compressed digital images, and involves first solving for the camera parameters and then solving for the point measurement data (often called "inverse" or "reverse" photogrammetric projects). A typical example is in the determination of the height of a suspect. Although there are manual methods using projective geometry to make the same determinations, an analytical photogrammetric method would first try to determine the parameters of the camera using known points of objects in the still image and then using this information to calculate the suspect's height.

It is also interesting to note that photogrammetry has no fixed accuracy since it is dependent on the camera capturing the image. Therefore, things such as camera resolution, lens distortions, and image quality all contribute to the resulting accuracy, and size of an object is not so relevant inasmuch as it is possible to physically capture a quality image of some object from varying angles. Therefore, with the right equipment, it is possible to measure minute details for small objects; on the other end of the spectrum, it is possible to take aerial images from an airplane or helicopter to map a large crime or accident scene.

5.3.2 Image Size and Measurement

The smallest element of a digital image is the pixel (picture element); the greater the number of pixels contained in an image, the greater the amount of resolution. An image that has a length and width of 4,592 × 3,056 pixels has an image resolution of 14,033,152 or approximately 14.0 megapixels. In terms of the number of measurements that can be captured from a digital photograph, there are a number of variables that can come into play, but fundamentally, measurements are directly related to the resolution of the image. Therefore, if two or more photos are taken of an area or object and the photos overlap by approximately 80%, it is possible to obtain

$$0.8\ \textit{overlap} \times 14{,}033{,}152 \text{ pixels} = 11{,}226{,}521 \text{ measurements.}$$

In fact, there are subpixel algorithms that allow for even finer measurements; however, these are interpolated measurements and not discrete measurements between pixels. There are also practical limitations to the quantity and quality of what is measured from photographs; in practice, they are usually based on factors such as how well the objects in the images are photographed or if they are clearly visible or obscured by another object (Figure 5.7). Often, a single point or set of distances is required to help relate

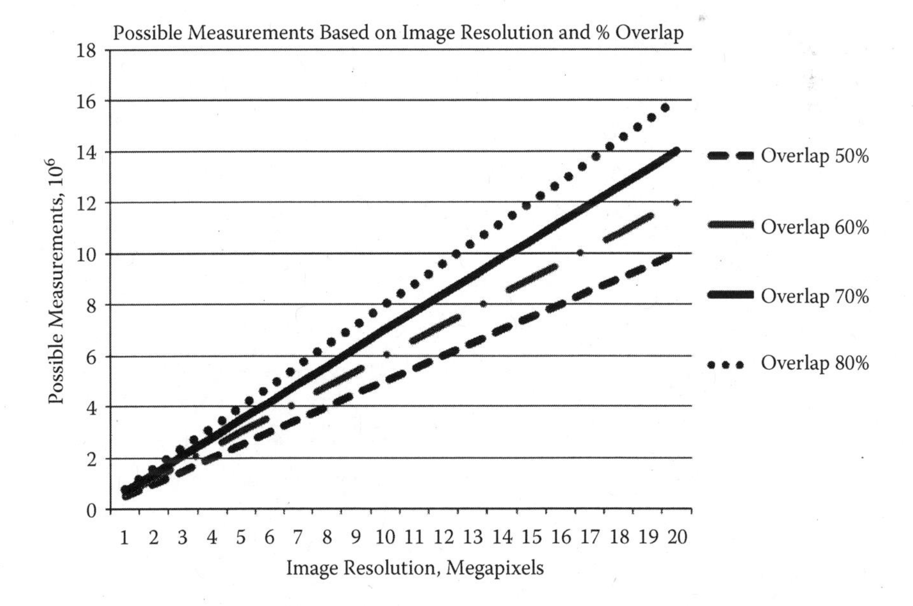

Figure 5.7 Possible measurements based on image resolution and percentage overlap.

the position or shape of evidence to the surrounding environment, and not all measurements for each pixel in an image are required.

5.3.3 Camera Distortions

Although most modern-day digital cameras are manufactured with reasonable quality, they are mostly used by the average person for taking portraits of family and friends and are less often considered for use in areas such as scientific research or industrial vision due to many internal imperfections. These imperfections contribute significantly to the way light is projected onto the image sensor and can be quantitatively determined when the camera is available. Once the properties of a particular camera are determined or are "known," we can use these properties to make corrections in a photogrammetric solution, and the camera is then considered to be "calibrated." Although there are some situations for which manual corrections can be made to lens distortions, they are normally not a simple thing to quantify unless many photographs are available of the same object viewed from wide angles of separation.

The most easily seen distortion is a lens aberration, for which there are two effects (Figure 5.8). The first is called barrel distortion and is the effect we see when using a wide-angle lens. This affects objects such as straight walls and doors so that they appear curved and bent outward like a barrel. A wide-angle, fish-eye lens is often used by photographers to give an artistic effect, but in photogrammetry, this is an undesirable result. The second type is called pincushion distortion and has the opposite effect, with objects looking pinched and straight lines curved inward.

In addition, there is a type of lens distortion called *decentering*. This type causes the optical center of an image to shift from the geometric center of the imaging plane due to any misalignment between the lens and the image sensor. The result of this distortion is that it shifts the imaging plane (left, right,

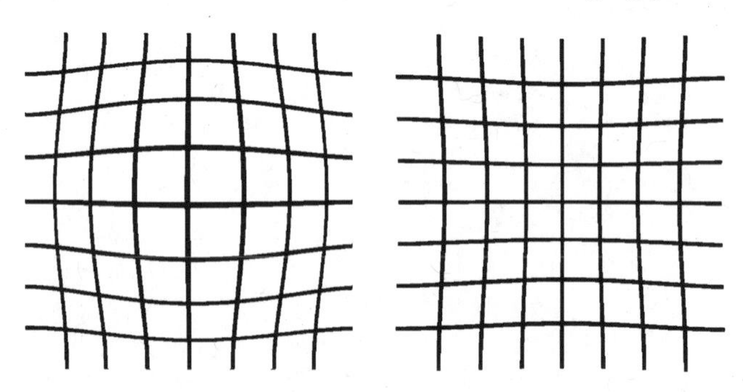

Figure 5.8 Barrel and pincushion distortion are the effect of lens aberrations.

up, or down) so that the center of the image (i.e., principal point) is located at a location other than the geometric center of the imaging plane.

5.4 Stereoscopic Images and Scanning Synopsis

One of the marvels of human vision is that humans use both eyes to process independent information about an object to determine how far away the object might be. Our depth perception is based on the interpretation by the brain of the left and right retinal disparities and can be thought of in part as the same effect as parallax between one eye and the other.

Stereo photogrammetry works in a similar manner and uses linear displacement (often referred to as the "base") between two cameras much like a set of aerial photographs from an airplane are displaced without changing the direction or "tilt" (in ideal situations) where the camera is looking (Figure 5.9). The amount of linear displacement between the two photographs will determine the optimum range of depth perception in a pair of images. The greater the base or displacement between cameras, the more accurate the depth calculations will be. However, this also means that there will be less

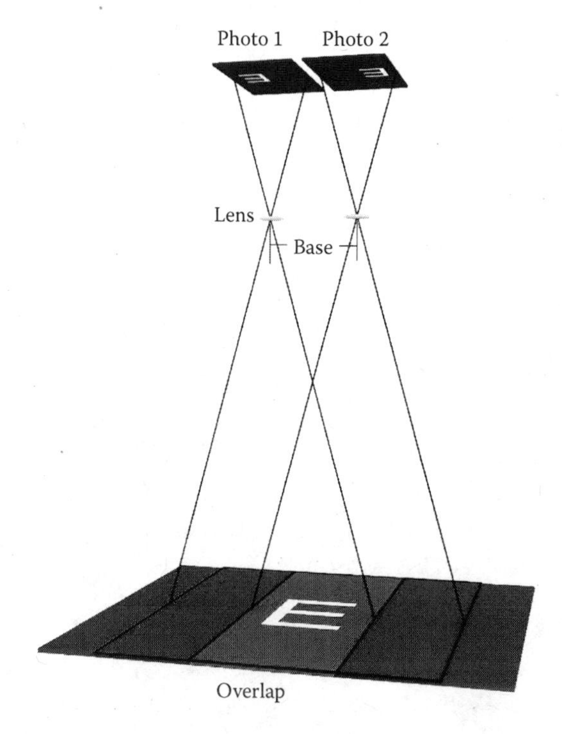

Figure 5.9 Simplified stereo pairs of photographs.

overlap in the images, and the resultant matching areas or actual point measurements that can be calculated are a direct result of the amount of overlap.

There are known relationships between the internal camera system and the external, real-world objects, and points can be calculated by geometric principles. However, the measurement calculations become increasingly complex when one takes into consideration all the small errors and deviations from an ideal system.

In modern-day photogrammetric systems that provide stereo photogrammetric solutions, the result starts with the creation of a 3D point where a pixel match has been made between photographs. There are matching algorithms that search for and match thousands of pixels; the result is a "cloud" of points representing the surface of the object.

Point clouds are best known as a product of 3D laser scanners and photogrammetric applications. They are normally used to represent the surface profiles of objects and at a minimum require the *x*, *y*, and *z* coordinates for any vertex or point measurement. Additional information, such as the color of the point given as values ranging from 0 to 255 for red, green, and blue (RGB), is often included. Different manufacturers of laser-scanning equipment often have their own proprietary format for point clouds, and sometimes they will include additional information such as surface normal, intensity, and other pertinent information.

The resultant point cloud (Figure 5.10) has several benefits in that the resulting data are normally dense. It is not uncommon for point clouds to number in the millions or even billions of points (Figure 5.11). Several point

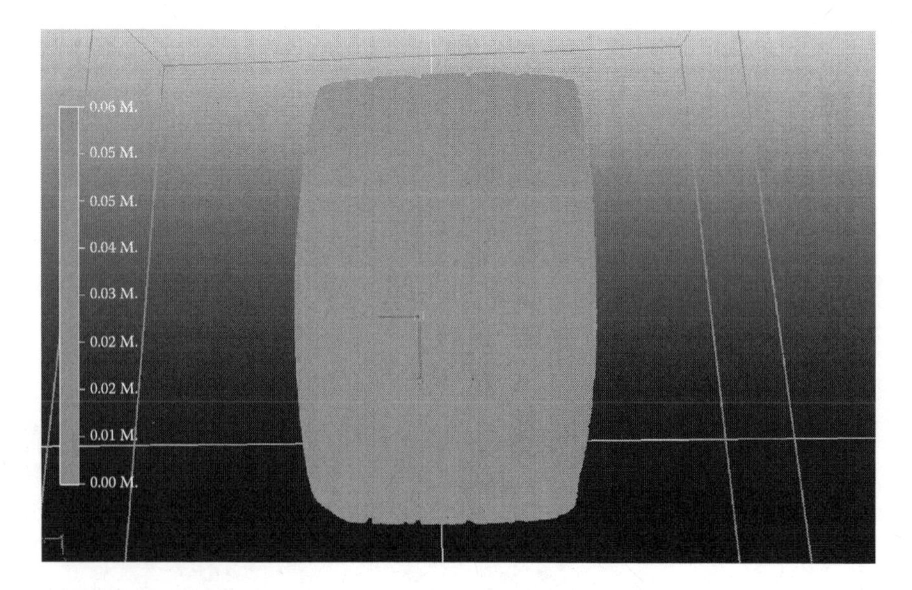

Figure 5.10 Point cloud for a portion of tire tread.

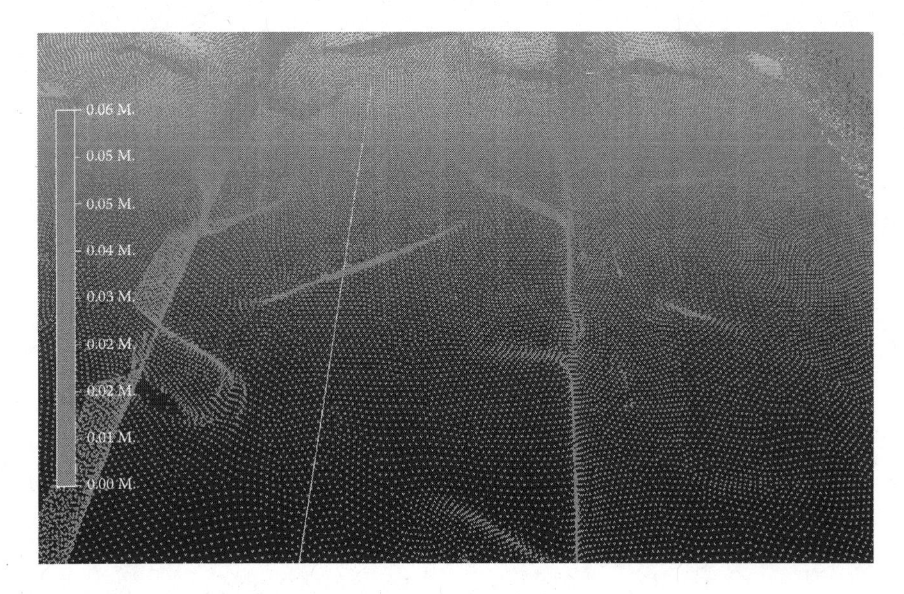

Figure 5.11 Close-up showing individual points in a portion of tread.

clouds can be automatically "stitched" using reference targets included in each scan or overlapping portions of each scan that can be recognized as common areas. *Registration* is the term often used for combining and accurately matching several independent or overlapping scans into one larger point cloud; there are now many software tools that aid in the analysis, processing, modeling, and visualization of point clouds.

References

1. Verma, S. 2008. *The little book of maths, theorems, theories and things*. New Holland Publishers, Chatswood, Australia, 163.
2. Ewen, D., and C. R. Nelson. 1995. *Elementary technical mathematics*. 6th ed. PWS Publishing, Boston, Maine, 175–176.
3. *McGraw-Hill concise encyclopedia of science and technology*. 2005. 5th ed. McGraw-Hill, New York, 2372–2374.
4. *McGraw-Hill concise encyclopedia of science and technology*. 2005. 5th ed. McGraw-Hill, New York, 689–690.
5. Center for Photogrammetric Training. *History of photogrammetry*. http://www.ferris.edu/faculty/burtchr/sure340/notes/HISTORY.pdf (accessed July 5, 2010).
6. Thompson, M. M. 1966. *Manual of photogrammetry*. 3rd ed., Vol. 1. American Society of Photogrammetry, Falls Church, VA.

Fluids

6

PATRICK MALLAY
DAVID S. PIERCE

The objective in this chapter is to offer some concise examples of studies in the role of fluids that in forensics is referred to as a matrix. A review of forensic literature devoted to fluids revealed considerable attention to bloodstain evidence. There is training that will introduce the topic of blood spatter, certification for blood spatter experts, and practically no mention of other fluids. This exclusion of information about other fluids amounts to a preoccupation with blood that is understandable on more than one level. Visual artists learn that composition is not only about lines or shapes but also even a small area of vivid color will attract and divert attention. No other color diverts attention as effectively as the color red. An example of blood bias is seen with presumptions that evidence must exist where the most obvious and grievous activities take place. It must be realized that most of the benefits derived from the study of bloodstains could also apply to studies of other fluids. There is a need to widen this forensic perspective to include studies of other common liquids and their behavior.

Bloodstains can provide reliable evidence regarding the individualization of a source (as in identification provided by DNA). Details about actions, the occurrence of multiple impacts, calculations of spatter velocities, or the distances between components of various acts are all derived from spatters and stains. Observations regarding voids in patterns can indicate the presence of persons or objects at a scene. Accumulations of such data can reveal the sequence of events and potentially contribute to an understanding of subsequent activities.

Human DNA is usually more relevant, yet the analysis of fluids other than blood can also offer compelling evidence. Samples of pond water, for example, can provide diatom evidence, and trace metals can be found in samples as small as a fingerprint. These examples reinforce the need to objectively consider the potential evidence of all fluids.

The idea that common liquids can also provide evidence suggests a need to expand on current understanding of the mechanisms and effects of a fluid matrix. The differences between some liquids are both visible and measurable. Those differences, as seen in this chapter, have been commonly ignored prior to this study.

6.1 Simple Drops

For the purposes of an experiment, *simple drops* were identified as drops of liquid released from an insignificantly small height and then allowed to dry. Pictures were taken of the drops once they were dry and after the dried drop was dusted with magnetic powder. These drops were released onto a variety of different surfaces so that effects such as adsorption, absorption, and other surface characteristics might be noted.

Liquids tested included chocolate milk, 1% milk, 2% milk, distilled water, oil, Coca-Cola©, Diet Coke©, Sprite©, double-double coffee, tea with milk, rubbing alcohol, vinegar, and filtered bottled water.

6.1.1 Apparatus

Liquids to be tested	Substrates to be tested	Magnetic fingerprint powder
Magnetic wand	Digital camera	Ruler
Thermometer	Cotton swabs	Lighting
Computer with photo-editing software		

6.1.2 Procedure

Perform the following steps in placing the drops:

1. Use a cotton swab to carefully place one drop of each liquid on the substrate.
2. Allow the drop of each liquid to dry.
3. Place the substrate in the light and take a picture of the dried drops.
4. Dust the surface of the substrate with magnetic powder.
5. Place the substrate in the light again and take a picture of the dusted drops.
6. Use photo-editing software to enhance the appearance of the drops.

6.1.3 Observations

In this experiment, an interesting observation that can be made in the photographs of Figure 6.1 is the existence of dark rings around the edges of many dried drops. Although some of these rings could be dismissed as simple clumps of powder, others must be attributed to some other cause as they can be seen in the photos of dried drops that were not dusted.

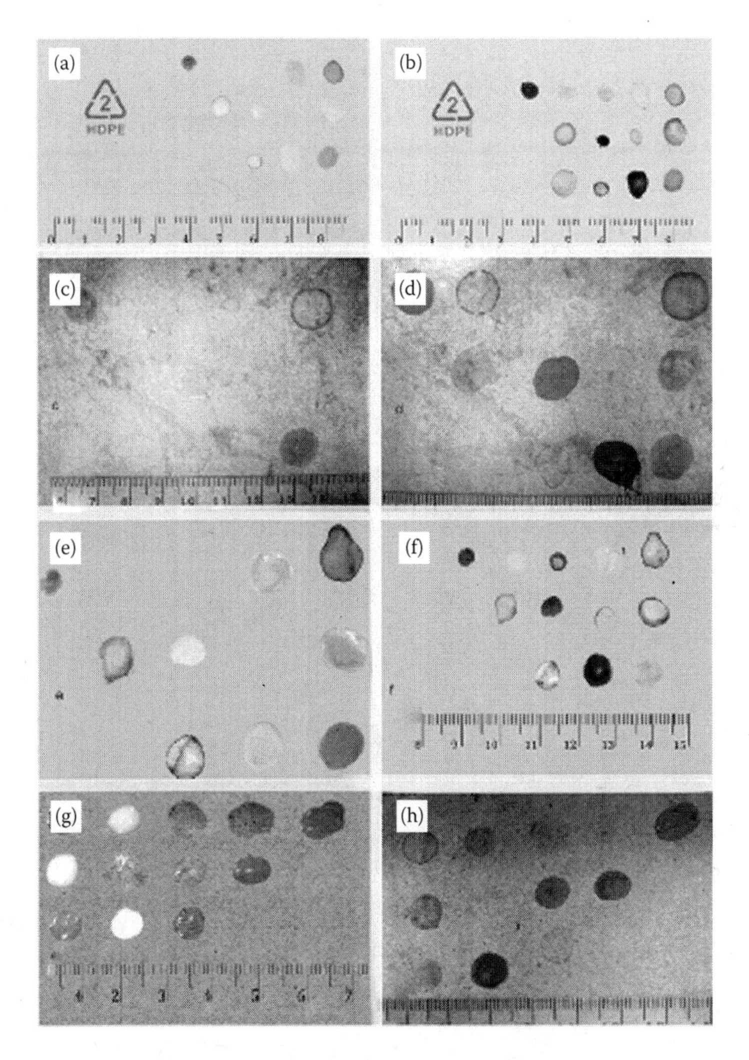

Figure 6.1 Simple drops. A comparison of photographs of both the dried drops and the dusted dried drops clearly shows that dusting can reveal the outlines of evaporated drops. This effect is most dramatic in side-by-side comparisons in which what appears to be "empty" space in (a), (c), (e), and (g) is shown to contain the remains of an evaporated drop or a lack of response in the corresponding dusted versions seen in (b), (d), (f), and (h).

The accepted explanation for this ring phenomenon was posited by Deegan et al.,[1] who explained that this behavior is caused by the contact line of the drop becoming pinned to irregularities in the substrate beneath it and subsequently forcing liquid within the evaporating drop to flow toward the edges (see Figure 6.2).

This outward flow is necessary for the drop to maintain the same area as its height shrinks. As the solvent evaporates from the edge of the drop, solute

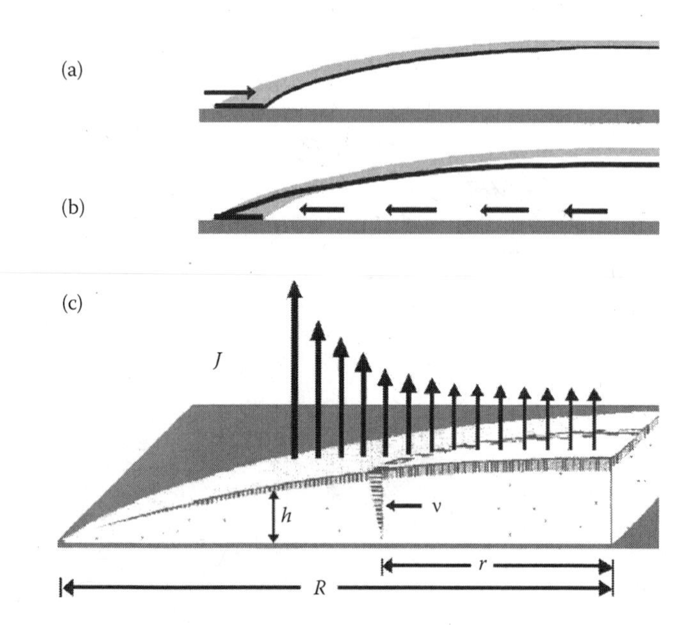

Figure 6.2 Schematic of capillary flow. "If the contact line is to remain in the same place, liquid must constantly flow outward to the edge of the drying drop. Capillary flow as the cause of ring stains from dried liquid drops." Mechanism of outward flow during evaporation. (a) and (b) show an increment of evaporation viewed in cross-section. (a) The result of evaporation without flow: the droplet shrinks. (b) The compensating flow needed to keep the contact line fixed. In (c) the quantities responsible for flow are defined. Vapour leaves at a rate per unit area $J(r)$. The removed liquid contracts the height $h(r)$ vertically, vacating the vertically striped region in a short time Dt. The volume of this striped region is equal to the volume removed by J. But in the shaded annular region the heavy-striped volume is smaller than the volume removed by J there (heavy arrows). Thus liquid flows outwards to supply the deficit volume: fluid at r sweeps out the horizontally striped region in time Dt. Its volume is the deficit volume; its depth-averaged speed is $v^{-}(r)$. J is the evaporating flux, v is flow velocity, h is the height of the droplet from the substrate (plane) to the cap as determined by surface tension, r is the radius vector (position), an R is the Ricc curvature tensor (droplet behavior on a plane surface). Reprinted by permission from Macmillan Publishers Ltd. Nature Publishing Group. Deegan, R. D., O. Bakajin, T. F. Dupont, G. Huber, S. R. Nagel, and T. A. Witten. 1997. *Nature*, 389(6653), 827–829, 1997.

is left behind to form a ring. Further development on this theory by Deegan et al.[2] resulted in an equation that relates the growth of the mass of the ring to time. As well, they identified that this equation is independent of the fluid, solids in the fluid, or substrate on which the drop is found.

6.1.4 Spattered Drops

In the experiment involving spattered drops, *spattered drops* were classified as drops of liquid released from a height necessary to ensure that they reached their terminal velocity prior to striking (and splattering) a target surface. To calculate the terminal velocity of a drop, it was necessary first to do a significant amount of research to determine the conditions under which a drop of liquid falls. Once a mathematical model had been created, testing was necessary to find experimental values for the variables that it required. Terminal velocities and the drop height necessary to reach those terminal velocities were then calculated for a number of different liquids using a computer program implementing the mathematical model. Of these liquids, drops of a selected few were tested and photographed both when dry and after dusting with magnetic powder. The images were then clarified with common graphic-editing software.

6.2 Modeling Droplets

6.2.1 Mathematical Model for the Volume of Liquid Droplets

The volume of a liquid droplet can be calculated using the pendant drop test, which equates the downward force of gravity on a drop to the upward force exerted by its surface tension. It is possible to do this because the instant before a drop falls from a tube (like a burette), these two forces will be equal and easy to calculate. The equation for this can be derived from that for a standard pendant drop test[3] and is as follows:

$$v = \frac{4\pi r \times s \times 10^{-4}}{g \times d}$$

where v is the volume (L), r is the radius of the burette tip (m), g is 9.81 m/s^2, d is the density (kg/L), and s is the surface tension of the liquid (dyn/cm).

6.2.2 Velocity of Liquid Droplets in Air

When a drop is falling through the air, two forces primarily influence its velocity. The force of gravity constantly acts downward on the drop, forcing it to accelerate toward the earth. At the same time, air resists the motion of the drop and provides a drag force that acts in an upward direction, opposing the force of gravity. The drag force varies depending on both the velocity and aerodynamic properties of the drop and at some point will be equivalent to the force of gravity. At this time, the drop will stop accelerating and will fall at a constant velocity. This is known as the terminal velocity of the drop.

The terminal velocity of a drop can be calculated using the formula

$$V_t = \sqrt{\frac{2(v \times d)g}{\rho\pi\left(\sqrt{\frac{4}{3}\pi 0.001v}\right)C_d}}$$

where V_t is the terminal velocity (m/s), v is the volume of the droplet (L), d is the density of the liquid (kg/L), g is 9.81 m/s², ρ is 1.2 kg/L, and C_d is the drag coefficient.

The equation that models the velocity of the drop over time is

$$v(t) = V_t^* \tanh\left(t \times \sqrt{\frac{g\rho C_d \pi \left(\sqrt{\frac{4}{3}} \pi 0.001 v \right)}{2vd}} \right)$$

where V_t is the terminal velocity (m/s), v is the volume of the droplet (L), d is the density of the liquid (kg/L), g is 9.81 m/s², ρ is 1.2 kg/L, C_d is the drag coefficient, and t is time (s).

We can then calculate the distance needed by the drop to reach its terminal velocity by evaluating the integral of its velocity from 0 until the time when its terminal velocity is reached.

6.3 Sample Calculation

For water dripping from a burette with r = 0.00054 m, d = 1 kg/L, and s = 71.97 dyn/cm,

$$v = \frac{4\pi(0.00054\ \text{m}) \times 71.97\ \text{dyn/cm} \times 10^{-4}}{9.81\ \text{m/s}^2 \times 1\ \text{kg/L}}$$

$$v = 50\ \mu\text{L}$$

Therefore, drips from the burette will contain 50 μL of water. This is within the generally accepted range for the volume of a drop.[4]

For a 50 μL drop with $v = 5 \times 10^{-5}$ L, d = 1 kg/L, and C_d = 0.47,

$$V_t = \sqrt{\frac{2\left(0.00005\ \text{L} \times \frac{1\ \text{kg}}{\text{L}}\right) 9.81\ \text{m/s}^2}{\frac{1.2\ \text{kg}}{\text{L}} \times \pi \left(\sqrt{\frac{4}{3}} \pi \times 0.00005\ \text{L} \times 0.0001 \right) \times 0.004}}$$

$$V_t = 3.77\ \text{m/s}$$

Therefore, the terminal velocity of each drop will be 3.77 m/s.

By simplifying the second half of the velocity-time equation, we can produce

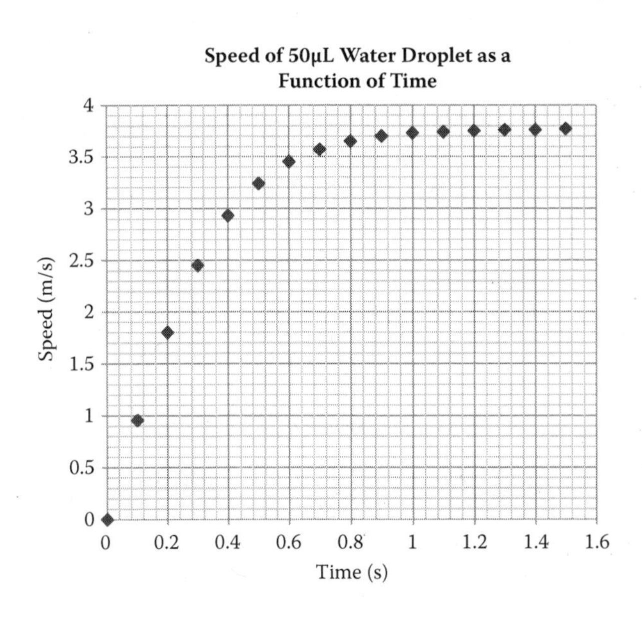

Time (s)	Velocity (m/s)
0	0
0.1	0.955
0.2	1.80
0.3	2.45
0.4	2.93
0.5	3.24
0.6	3.45
0.7	3.57
0.8	3.65
0.9	3.70
1	3.73
1.1	3.74
1.2	3.75
1.3	3.76
1.4	3.76
1.5	3.77

Figure 6.3 Water droplet speed. Table and plot illustrating the speed of a 50-μL water droplet as a function of time. As can be seen from the graph (left) and the table (right), the water droplet reaches its terminal velocity after 1.5 seconds. This is consistent with results obtained experimentally.

$$V(t) = 3.77 \tanh(2.59t)$$

which we use to produce the table and graph in Figure 6.3.[5]

To determine the distance that the droplet must fall to reach its terminal velocity, we take

$$\Delta d = \int_0^{1.5s} 3.77 \tanh(2.59t)\,dt$$

Noting that

$$\int \tanh(ax)dx = \frac{1}{a}\ln(\cosh(ax)) + C$$

This evaluates to

$$\Delta d = 1.4556 \ln(\cosh(2.59 \times 1.5s))$$

$$\Delta d = 4.65 \text{ m}$$

This shows that the height required by a 50-μL water droplet to accelerate to its terminal velocity is 4.65 m.

6.4 Experimental Process

6.4.1 Apparatus

Liquids to be tested	Substrates to be tested	Magnetic powder
Magnetic wand	Digital camera	Ruler
Thermometer	Cotton swabs	Lighting
Digital balance	Weighing bottle	Beaker
Computer with photo-editing software		

6.4.2 Procedures

6.4.2.1 Determining the Surface Tension and Density of Test Liquids

1. Mount the clean and dry burette on the vertical stand.
2. Determine the mass of the beaker.
3. Find the temperature in the laboratory.
4. Fill the burette with 10 mL of distilled water.
5. Allow 20 drops of water to drip from the burette into the beaker.
6. Find the mass of the beaker with water and determine the mass of 20 drops.
7. Empty the beaker and the burette. Clean and dry them in preparation for the next measurement.
8. Repeat steps 3 to 7 for all liquids.
9. Calculate the density of each liquid in kilograms per liter.
10. Take the temperature in the laboratory, determine the water surface tension (dyn/cm^2) using values from Table 6.1, and calculate the surface tensions of studied liquids according to the relation

$$\sigma = \sigma_{H_2O} \times \frac{m}{m_{H_2O}}$$

Surface tension was determined using a method adapted from the experiment given in "The Surface Tension of Liquids Measured with the Stalagmometer," designed by R N Dr. J. Gallová and translated by N. Kučerka from the faculty of pharmacy at Comenius University in Bratislava, Slovakia. The original experiment can be found at http://www.fpharm.uniba.sk/.../The_surface_tension_of_liquids_measured_with_the_stalagmometer.pdf.

6.4.2.1.1 Observations Table 6.2 presents the results for the determination of the surface tension and density of test liquids.

Table 6.1 Density and Surface Tension of Distilled Water at Given Temperatures

t (°C)	σ(dyn/cm²)	ρH_2O (kg/L)
15	73.49	0.99996
16	73.34	0.99994
17	73.19	0.99990
18	73.05	0.99985
19	72.90	0.99978
20	72.75	0.99820
21	72.59	0.99799
22	72.44	0.99777
23	72.28	0.99754
24	72.13	0.99730
25	71.97	0.99705

Table 6.2 Determining the Surface Tension and Density of Test Liquids

Liquid	Mass per 20 Drops (g)	Temp (°C)	Surface Tension H_2O (dyn/cm²)	m H_2O (g)	Surface Tension (dyn/cm²)	Density (kg/L)
Tap water	1.34	21.9	72.44	1.35	71.9	0.995
Distilled water	1.35	23.0	72.28	1.35	72.3	0.999
Dasani™	1.03	23.0	72.28	1.35	55.1	0.978
Low-concentration brine (10% NaCl)	1.36	21.6	72.44	1.35	73.0	1.06
High-concentration brine (26% NaCl)	1.42	21.6	72.44	1.35	76.2	1.18
Vinegar (5% CH_3COOH)	0.73	23.0	72.28	1.35	39.1	1.01
2% milk	0.58	23.0	72.28	1.35	31.1	1.01
Glycerol	0.95	23.0	72.28	1.35	50.9	1.03
10% cream	0.86	21.6	72.44	1.35	46.1	1.03
Skim milk	0.76	23.0	72.28	1.35	40.7	0.995
Chocolate milk	0.94	23.0	72.28	1.35	50.3	1.05
Black coffee	1.02	21.6	72.44	1.35	54.7	1.01
Double-double coffee	0.98	21.6	72.44	1.35	52.6	0.966
Black tea	1.26	21.6	72.44	1.35	67.6	1.03
Tea with 2% milk	1.15	21.6	72.44	1.35	61.7	0.996
Rubbing alcohol	0.31	23.0	72.28	1.35	16.6	0.867
Coca-Cola	0.97	23.0	72.28	1.35	51.9	0.949
Diet Coca-Cola	0.92	23.0	72.28	1.35	49.3	1.04
Citrus pop	1.32	21.6	72.44	1.35	70.8	1.09
Hydrogen peroxide	1.14	21.6	72.44	1.35	61.2	1
SAE 10W-40 Oil	0.59	21.9	72.44	1.35	31.7	0.9

Table 6.3 Determining the Drop Height

Liquid	Drop Volume (μL)	Terminal Velocity (m/s)	Drop Height (m)	Time (s)
Tap water	23	3.1	3.2	1.2
Distilled water	23	3.1	3.2	1.2
Dasani™	18	2.9	3	1.2
Low-concentration brine (10% NaCl)	22	3.2	3.3	1.3
High-concentration brine (26% NaCl)	21	3.3	3.6	1.3
Vinegar (5% CH_3COOH)	12	2.7	3	1.3
2% milk	9.9	2.5	2.9	1.3
Glycerol	16	2.9	3.2	1.3
10% cream	14	2.8	3.1	1.3
Skim milk	13	2.7	3	1.3
Chocolate milk	15	2.9	3.2	1.3
Double-double coffee	17	2.9	3.2	1.3
Tea with 2% milk	20	3	3.3	1.3
Rubbing alcohol	6.1	2.1	2.4	1.3
Coca-Cola	18	2.8	3.2	1.3
Diet Coca-Cola	15	2.9	3.2	1.3
Citrus pop	21	3.2	3.5	1.3
Hydrogen peroxide	20	3	3.3	1.3

6.4.2.2 *Determining the Drop Height*

1. Open DropHeightCalculator.exe (see Appendix) and enter the experimental values for density and surface tension determined in Section 6.4.2.1.
2. Record the returned values for drop volume, terminal velocity, and drop height for future reference.

6.4.2.2.1 Observations Table 6.3 presents the determinations obtained for the drop height.

6.4.2.3 *Performing the Drop*

1. Fill the burette with a tiny amount of the liquid to be dropped.
2. Draw circles on the substrate material (one for each liquid to be dropped) with the fine-tip marker.
3. Place the burette at an altitude sufficient for the drop of liquid to accelerate to its terminal velocity and aligned so that the drop will fall in an empty circle.

4. Open the valve on the burette only enough for one drop to fall out of the burette.
5. Using the digital camera, record the splattered drop.
6. Clean and dry the burette; repeat steps 1–5 for each material.
7. One to two days later, record all splattered drops again.
8. Dust all spattered drops and record again. Manipulate with photo-editing software to achieve the clearest patterns.

6.4.2.3.1 Observations In the previous mathematical mode, there is somewhat of a paradox presented to anyone who wishes to determine either the height from which a drop fell or the volume it contained by examining the dried stain (Figure 6.4). A straightforward explanation was provided in *Crime Scene to Court: The Essentials of Forensic Science*:[6]

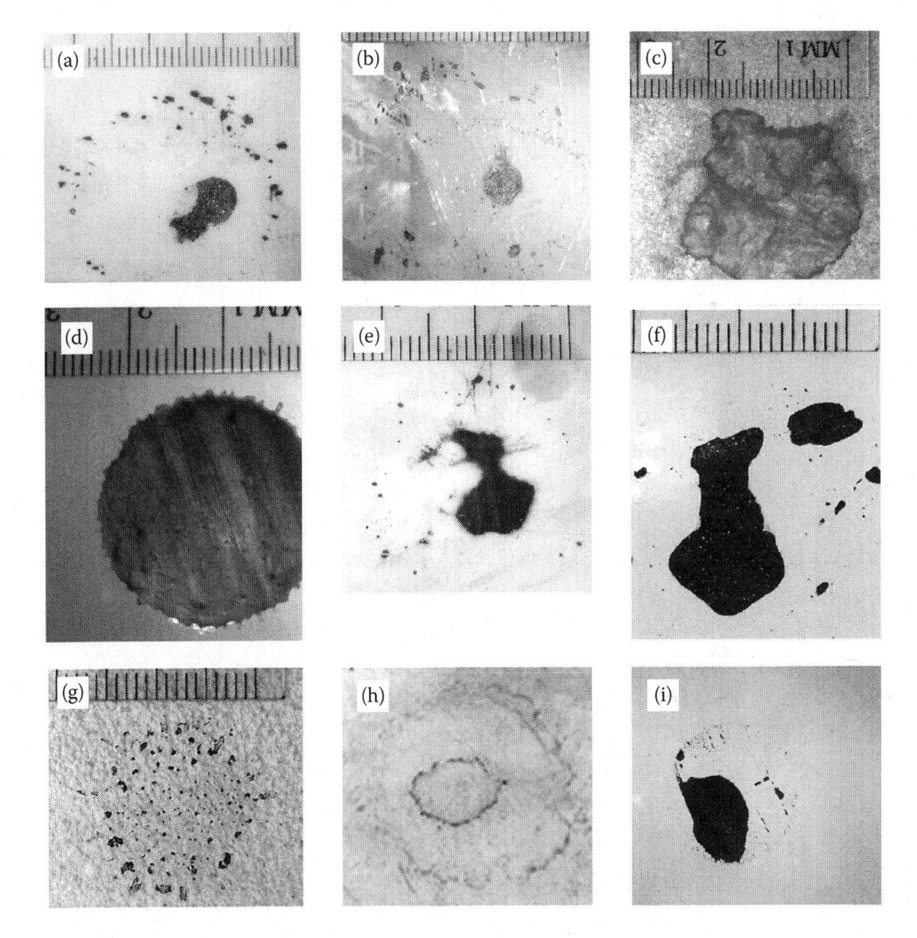

Figure 6.4 Observations of dropped liquids.

> The size and shape of the stain produced by a drop of blood falling under the influence of gravity perpendicular onto a target surface will depend upon three variables:
>
> 1. The volume of the individual drop—the greater the volume the larger the stain for the same dropping height and the same target surface.
> 2. The dropping height—increasing height produces an increase in stain size until the terminal velocity is reached, which is approximately 7 m for a 50 μl drop.
> 3. The nature of the target surface—non-absorbent surfaces will lead to the formation of larger stains as the entire volume of the drop spreads over the surface, whereas absorbent surfaces will tend to produce smaller stains as a proportion of the blood is absorbed within the substrate.
>
> In general terms, therefore, the size and shape of a stain can give *no* [italics added] indication of the dropping height unless the other variable parameters of the drop's volume and the effect of the target surface is known. (p. 118)

Although the scenario described by White is correct if a drop of any liquid is looked at from the traditional mechanical engineering perspective, there is fortunately another recently discovered method that may enable forensic investigators to back-calculate the height from which a drop fell.

In Figure 6.4e, a close examination of the dried drop reveals a number of "spines" protruding from the stain. Research[7] has been conducted that demonstrated how the impact velocity of a dried drop could be determined by counting these spines. This takes into account an experimentally determined "spread factor" for the impacted surface along with the Weber and Reynolds numbers of the drop liquid. In a blind trial evaluation of their method 2 years later,[8] these authors confirmed their original findings and revealed that the technique now allows calculations that are accurate to within 3.7 m/s (impact velocity) and 1.7 μL (drop volume). Despite these encouraging findings, Hulse-Smith et al. were careful to note that more experiments are necessary before their technique is ready for casework implementation.

6.5 Friction and Fluids

The purpose of this experiment was to explore how the friction between a rubber shoe and a substrate is affected by the addition of different liquids between the two contacting surfaces. Tests were conducted using a British-style pendulum tester to simulate the motion of a foot striking the ground, so the results might be considered applicable to a human attempting to run on a liquid-covered surface. When developing the experiment, it was assumed that friction would be the only force affecting the amount of time that the pendulum took to stop.

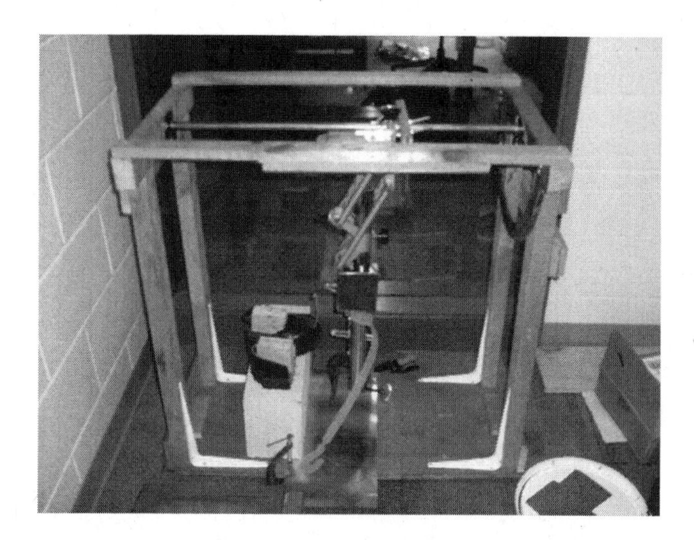

Figure 6.5 The British-style pendulum friction tester used in this experiment.

6.5.1 Apparatus

The apparatus for this experiment were as follows: British-style pendulum tester (see Figure 6.5), liquids to be tested, substrates to be tested, footwear to be tested, and a stopwatch.

6.5.2 Procedure for Determining Effect of Liquids on Friction

1. Place the substrate to be tested under the British-style pendulum tester.
2. Attach the footwear to the arm of the tester so that it touches the surface of the substrate when at rest.
3. Release the pendulum tester once to check that the footwear just glances the substrate when it goes through a full swing. If the footwear seems to have too heavy an impact on the substrate, then shorten the pendulum arm.
4. Release the pendulum tester again. This time, start the stopwatch when the tester is released and record the amount of time that passes until the tester stops moving.

Perform step 4 ten times, making sure that the surface retains a constant coating of liquid and the pendulum is released from the same height during each trial.

6.5.2.1 Observations

Observations from the experiment on friction and fluids can be found in Figures 6.6 to 6.8.

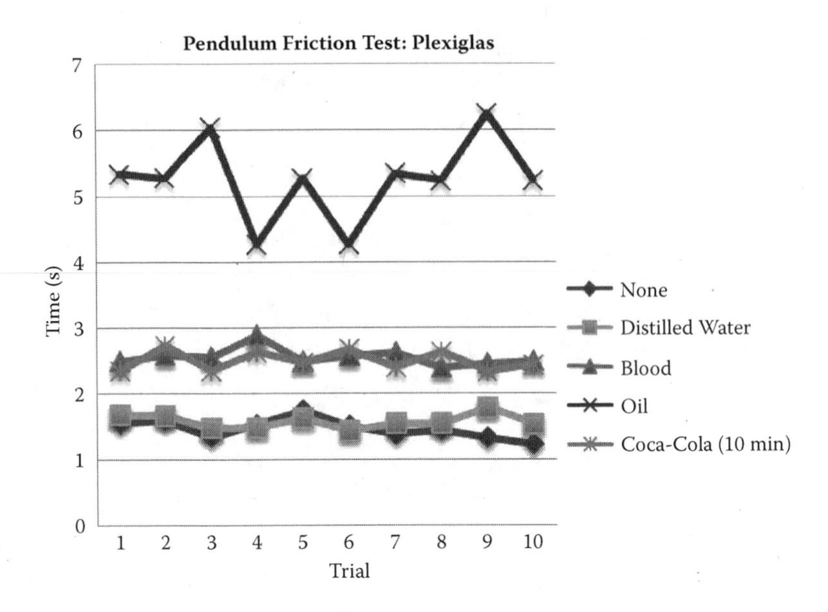

Figure 6.6 Results of a pendulum test between a rubber shoe and liquid-coated Plexiglas™.

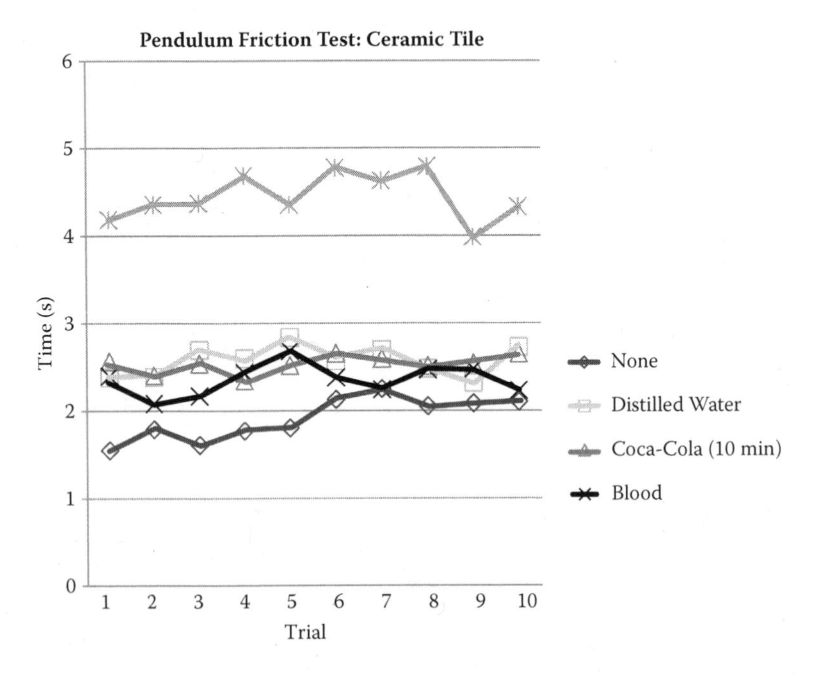

Figure 6.7 Results of a pendulum test between a rubber shoe and liquid-coated ceramic tile.

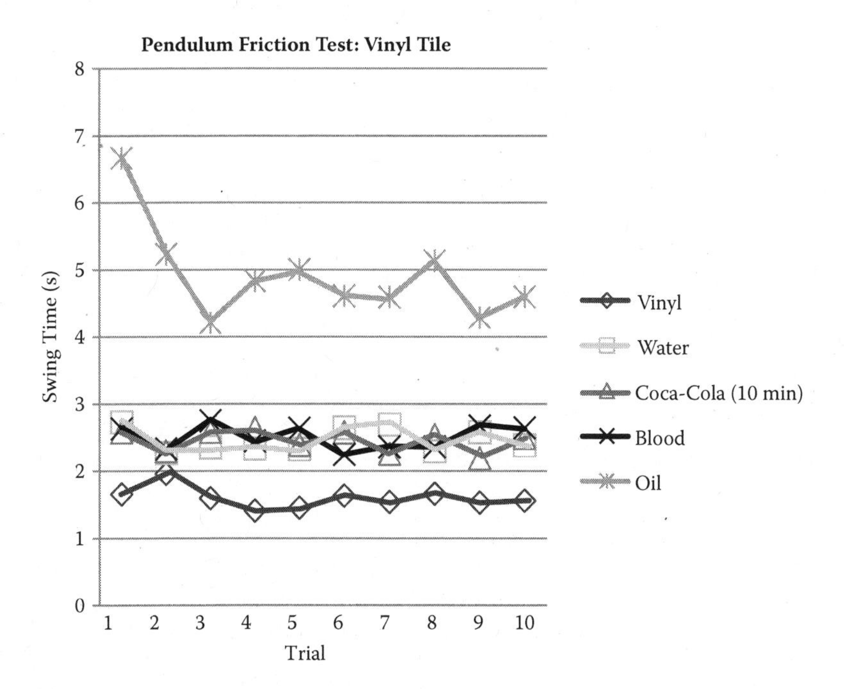

Figure 6.8 Results of a pendulum test between a rubber shoe and liquid-coated vinyl tile.

6.5.2.2 *Discussion*

From the data collected during this experiment, it appears that the effect liquids have on reducing friction is independent of the surface on which they reside. This can be concluded by examining Figures 6.6 to 6.8 and noting that although the amount of time necessary to stop the pendulum varied between surfaces, each liquid had a similar effect on each surface on which it was tested. For example, oil significantly decreased the friction (increased the time to stop) on ceramic tile, vinyl tile, and Plexiglas. Similarly, Coca-Cola slightly decreased the friction on each surface.

6.5.3 Rubber Friction

A common misunderstanding that many professionals share is the assumption that models and equations used to model friction between metal surfaces can also be applied to model friction between a rubber surface and a metal one. This is because, unlike the constant coefficient of friction that occurs between metal surfaces, rubber coefficients of friction are dependent on the force applied to the rubber as it is moved.

The main reason for the changing coefficient of friction for rubber is that the force-producing mechanism in rubber is elastic deformation, while in metal it is plastic deformation. As the load on the rubber increases, there is less resistance to bulk deformation, leading to a lower coefficient of friction.[9]

6.5.4 Matters Involving Blood

Blood not only draws attention but also causes a knee-jerk reaction. There is a primordial sensation of danger in the presence of blood at a crime scene that can fixate the best of us[10]:

> Exposed human blood is not unlike other common fluids. It will act in a predictable manner when subjected to external forces. Blood, whether a single droplet or a large volume, is held together by strong cohesive molecular forces that produce a surface tension within each drop. Surface tension is defined as the force that pulls the surface molecules of a liquid toward its interior decreasing the surface area and causing the liquid to resist penetration. The surface tension of blood (0.058 γ(N/m) at 37°C) is close to the surface tension of water (0.073 γ(N/m) at 20°C and 0.059 γ(N/m) at 100°C).

Blood is a colloid, a liquid base of serum that contains dispersions of other materials. It is entirely possible that the mechanisms, for both the behavior and drying properties, differ in some significant ways from those of other common liquids. While blood has been the subject of some study in terms of its droplet, spatter, and pooled effects, one should remember that other liquids may not behave in the same way. The behavior of liquids aside from blood is rarely mentioned and has not previously been considered as a source of evidential merit.[11]

6.6 Analysis and Comparison

6.6.1 The Value of Droplet Studies to Analysis and Comparison

An explanation was introduced in this chapter (capillary action) to explain the appearance of a ringlike drying effect with some liquids. This model does not entirely explain the appearance of droplets that do not seem to exhibit a capillary drying response, such as some thick bloodstains or paint.

The difference between drying characteristics is illustrated by the droplets shown in Figures 6.9 and 6.10.

While both cream and blood are colloids, in diluted form they behave similarly to less-dense substances. In relatively viscous form, they exhibit capillary action, yet these substances do not regularly form the ring pattern seen with more dilute examples. The reasons for these disparate drying characteristics may include coagulation or flocculation.

The exact causes and identification of the mechanisms involved in determining the point at which a solution dries as a solid may prove interesting

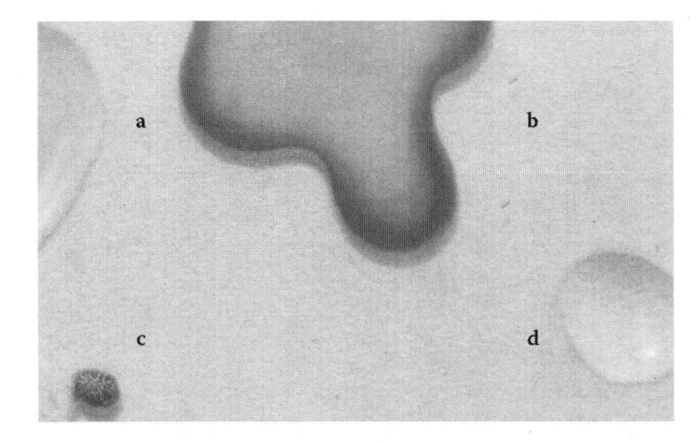

Figure 6.9 Capillary drying experiment. The differences in drying can be accounted for by the nature and structure of the droplets involved. This simple experiment consisted of four droplets on glass substrate. The droplets were as labeled: (a) diluted blood (a 10% solute of the author's blood mixed with distilled water); (b) coffee; (c) undiluted blood; (d) a droplet of tap water.

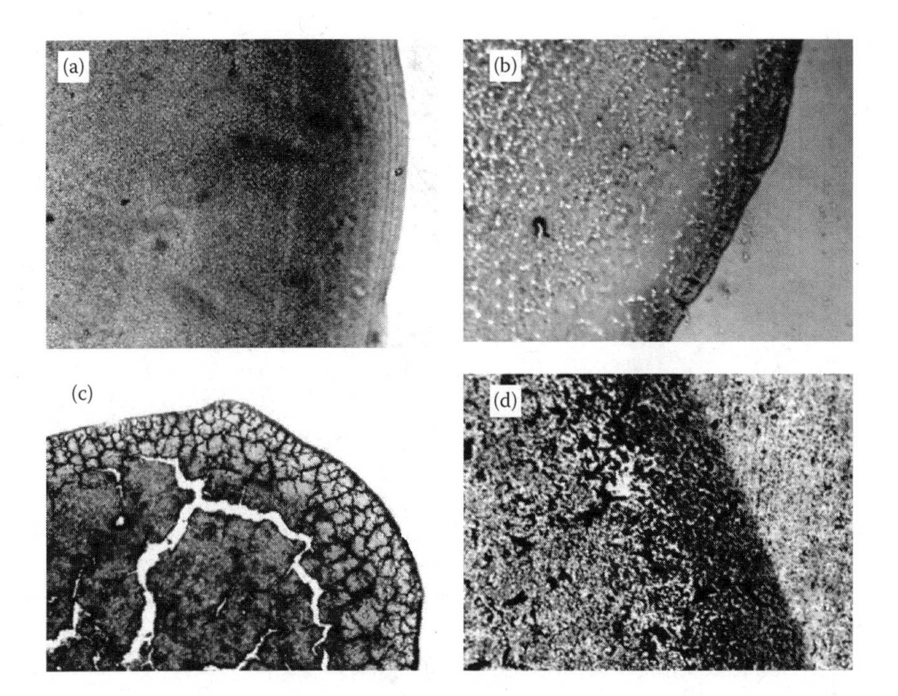

Figure 6.10 Microscopic enlargement of capillary effects. The enlarged versions (ranging between ×80 and ×160) (derived from the experimental conditions shown in Figure 6.9) include (a) dried 10% dilution of blood in distilled water; (b) dried coffee drop; (c) dried blood droplet; and (d) a water droplet (only this specimen was dusted with black magnetic powder).

for future research. While most familiar colloids seem to behave in much the same way, it is the subtle differences that allow discrimination of their effects. Certainly, it can be said that a colloid dispersion capable of leaving a ring stain can be predicted to have contributed a liquid effect to any contact where the effect is observed. Table 6.4 shows information for some typical colloidal systems.

The next logical ideal step is to apply mathematics. Comparisons could utilize a variety of nodal points of correspondence. The problem with this approach is the difficulty in expressing assessments of relative clarity and sufficiency in a mathematical format.

It would take a large software program indeed to achieve the subtlety of making a relative assessment of a marking. Friction ridges are flexible and will often require a great deal of interpretation. Tool, tire, and footwear markings may also appear different in size, shape, or apparent pattern yet derive from the same source.

The possibility looms larger with each leap of technology that a computer will one day have completed almost all if not all of an analysis and comparison (including statistical verification of its own results). The use of computers to analyze evidence began decades ago and will continue to evolve. One has to wonder, though, if the transition will live up to expectations, just as computers were once considered poised to lead us into a world without reliance on paper.

Consider influences such as variances in pressure, adsorption, hysteresis, or dimensional stability, which may seem simplistic taken separately. Combinations of effects may behave in a synergistic manner. Imagine a marking that has been altered from what may be considered "normal" or "typical." Can you account for any differences in the appearance? Is the value of your opinion diminished without an adequate explanation for any differences?

6.6.2 Capillarity

We have seen that liquids are bound by cohesive forces and that a liquid will exhibit some amount of attraction to a solid surface. Fluids with a contact angle θ less than 90° will tend to "wet" the surface they are on (thereby exhibiting a stronger attraction to the molecules of a solid surface than to other molecules of the same fluid). This means that the adhesive forces are stronger than the cohesive forces at this particular liquid-to-solid interface.

For wetting to occur, the solid-to-gas interface (existing condition usually involving the substrate in air) must possess a strong attraction. The solid-to-gas interface must exhibit a surface energy that exceeds the combined surface

Table 6.4 Some Typical Colloidal Systems

Examples	Class	Nature of the Disperse Phase	Nature of the Disperse Medium
	Disperse Systems		
Fogs, mists, tobacco smoke, "aerosol" sprays	Liquid aerosol or aerosol of liquid particles[a]	Liquid	Gas
Industrial smokes	Solid aerosol or aerosol of solid particles[a]	Solid	Gas
Milk, butter, mayonnaise, asphalt, pharmaceutical creams	Emulsions	Liquid	Liquid
Inorganic colloids (gold, silver, iodide, sulfur, metallic hydroxides, etc.), paints[b]	Sols or colloidal suspensions	Solid	Liquid
Clay slurries, toothpaste, muds, polymer lattices	When very concentrated, called a paste	Solid	Liquid
Opal, pearl, stained glass, pigmented plastics	Solid suspension or dispersion	Solid	Solid
Froths, foams	Foam[c]	Gas	Liquid
Meerschaum, expanded plastics	Solid foam	Gas	Solid
Microporous oxides, "silica gel," porous glass, microporous carbons, zeolites	Xerogels[d]		
	Macromolecular Colloids		
Jellies, glue	Gels	Macromolecules	Solvent
	Association Colloids		
Soap/water, detergent/dye solutions	—	Micelles	Solvent
	Biocolloids		
Blood		Corpuscles	Serum
Bone		Hydroxyapatite	Collagen
Muscle, cell membranes		Protein structures, thin films of lethecin, etc.	

(continued on next page)

Table 6.4 (continued) Some Typical Colloidal Systems

		Nature of the	
Examples	Class	Disperse Phase	Disperse Medium
Three-Phase Colloidal Systems (Multiple Colloids)		Coexisting Phases	
Oil-bearing rock	Porous rock	Oil	Water
Capillary condensed vapors	Porous solid	Liquid	Vapor
Frost heaving	Porous rock or soil	Ice	Water
Mineral flotation	Mineral	Water	Air bubbles or oil droplets
Double emulsions	Oil	Aqueous phase	Water

Source: *With permission Everett, D. "Basic Principles of Colloid Science." Chapter 1, pp. 4–5, 1988, Royal Society of Chemistry.*

[a] Preferred nomenclature according to IUPAC (International Union of Pure and Applied Chemistry) recommendations.

[b] Many modern paints are more complex, containing both dispersed pigment and emulsion droplets.

[c] In a foam, it is usually the thickness of the dispersion medium film that is of colloidal dimensions, although the dispersed phase may also be finely divided.

[d] In some cases, both phases are continuous, forming interpenetrating networks, both of which have colloidal dimensions.

energies of the liquid-to-gas attractions and the solid-to-liquid attractions. With samples that exhibit stronger cohesion forces at interfaces than the solid-to-gas interface, the liquid will "bead up" on the surface.

Capillarity is the other phenomenon that affects the thresholds of vision in the collection or even location of liquids at a crime scene. Fluids can be lifted to great heights, depths, or recesses by capillary action. Capillary action is demonstrated by fabrics that allow fluids to "wick" or seep through unless treated. Waterproofing of the same fabric chemically changes that fabric to a wetting-resistant material on which most liquids will bead on the surface.[12]

Conditions necessary for capillary action to take effect include that the adhesion to a surface must be stronger than the cohesive forces between the molecules of the liquid. This is nicely demonstrated by the movement of liquid wax as it moves up the wick of a burning candle. Capillary actions are defined in calculation by the component liquid-to-gas surface tension, density of the liquid, contact angle, and size of the fissure or tube opening.

The capillary force acting on a penetrant that is driven into a crack or fissure may be expressed as follows:

$$\text{Force} = 2\pi\pi\sigma_{LG} \cos\theta$$

where the radius of the crack or fissure opening is represented by π, the liquid-gas surface tension by σ_{LG}, and the contact angle as θ.[13]

The force in this instance is part of the greater calculation of capillary pressure. There are other considerations, depending on the type of capillary action in question, the density and surface tension of the liquid, and the pore or opening size of materials involved.

Adhesion tension is, at times, the active force causing penetration of a fluid into a flaw. No matter the origin of the force, capillary pressure will cause a liquid to continue filling the void unless some opposing force, such as the building pressure of a trapped gas (unable to escape the influx of liquid within the covered flaw), changes the balance of forces at work.

Surfaces that are contaminated with oil, grease, water, or other substances prior to exposure may impede or ruin any evidence. An exception is the contaminant as the evidence, such as with footwear imprints. It can also be possible in the matter of a highly machined substrate, such as the ground surface of a metal safe, for trace evidence to have become encapsulated inside the ground area.

Expectations that evidence has been degraded by cleaning with a solvent (especially bleach in biological examinations), rinsing with water, or even applications of an emulsifier are well founded. Such treatment will have a dramatic effect on the surface. The treatment may also have eliminated evidence within the flaw. There is a possibility that treatment with a developer in the form of a dry powder, wet-dipped solution, or sprayed application could draw the evidence back to the surface. This is essentially the hoped result with use of a luminescent reagent such as a luminal derivative.

A chemical reagent that exhibits chemifluorescence with exposure to a suitable oxidizing agent and in the presence of a catalyst is used by biologists to detect copper, iron, and cyanides. In a laboratory, potassium ferricyanide may be used as the catalyst, and at a crime scene, iron compounds usually perform that function.

Fluorescing reagents must be photographed. Markings do not always appear in a location that can be photographed; in those instances, other techniques may be employed. An alternative technique is the use of a "blind lift," such as used on the door handle of a vehicle or the interior of a machine that has been pried. These are cases that do not warrant dismantling the door or the device to treat an obscured substrate where markings can possibly be located using capillary action.

6.6.3 Rough Surfaces

Material science teaches us that even smooth materials are not microscopically smooth if larger flaws are visible; other surface conditions exist as barely visible or relatively invisible features. Fractures or microfissures that warn of strength failure problems can also relate to impression evidence by trapping minute traces of evidential value.

Microroughness and flaws are common on apparently smooth objects, including glass, metals, many ceramics, rubber, and plastics. The types of flaws that may be expected on some of these materials include fatigue cracks, quench cracks, grinding cracks, overload and impact fractures, porosity, laps, seams, pinholes in welds, and fusion or braising artifacts at the edge of a bond line. While other substances (e.g., epithelial cells or foreign molecules) could be found, fluids may well be present in these materials and circumstances.

6.7 Summary

Portions of this study made only brief mention of important surface conditions and effects, including adsorption, absorption, drying characteristics, and the coefficient of friction for rubber. These topics will benefit greatly from additional examination in other chapters.

Differences were noted in the appearance of droplets on drying, both before (in some instances) and after treatment with magnetic powder.

Spatter differences within the body of research presented many similar traits to blood spatter, including the formation of spines in spatter resulting from near-perpendicular test conditions. This result was confirmed independently (in direct discussion with Lee Smith, interviewed by D. S. Pierce, October 14, 2009) as consistent with testing conducted using colored water. These experiments preceded the published work.[7]

A direct benefit of the research is the potential (in some cases) to determine the threshold values at which a droplet may or may not appear as found on a particular substrate.

The friction studies confirmed that

- The presence of a particular matrix is likely a primary factor in considering friction effects on various types of flooring.
- Consideration of elastic friction with rubber outsoles cannot be expected to conform to the plastic deformations encountered with metals.

Throughout the experiments summarized in this chapter, interesting phenomena from all areas of science were encountered. The usefulness of each technique will depend on more factors than can be guessed here. For the curious forensic practitioner, there is virtually unlimited potential for further experimentation and research into these and other previously neglected areas.

References

1. Deegan, R. D., O. Bakajin, T. F. Dupont, G. Huber, S. R. Nagel, T. A. Witten. 1997. Capillary flow as the cause of ring stains from dried liquid drops. *Nature*, 389, 827–829.
2. Deegan, R. D., O. Bakajin, T. F. Dupont, G. Huber, S. R. Nagel, and T. A. Witten. 2000. Contact line deposits in an evaporating drop. *Physics Review E*, 62(1), 756–765.
3. Various authors. 2009. Drop (liquid). Wikipedia. http://en.wikipedia.org/wiki/Drop_(liquid).
4. Van Mook, F. J. R. 2002. *Driving rain.* Bouwstenen series of the Faculty of Architecture, Planning and Building of the Eindhoven University of Technology. Issue 69. http://sts.bwk.tue.nl/drivingrain/fjrvanmook2002/node8.htm.
5. Gunn, R., and G. D. Kinzer. 1949. The terminal velocity of fall for water droplets in stagnant air. 1949. *Journal of Meteorology.* http://staff.science.uva.nl/~jboxel/Publications/PDFs/ Gent_98.pdf.
6. White, P. 2004. *Crime scene to court: The essentials of forensic science.* RSC.
7. Hulse-Smith, L., N. Z. Mehdizadeh, and S. Chandra. 2005. Deducing drop size and impact velocity from circular bloodstains. *Journal of Forensic Science*, 50(1), 1–10.
8. Hulse-Smith, L., and M. Illes. 2007. A blind trial evaluation of a crime scene methodology for deducing impact velocity and droplet size from circular bloodstains. *Journal of Forensic Science*, 52(1), 68–69.
9. Smith, R. H. 2008. *Analyzing friction in the design of rubber products and their paired surfaces.* CRC Press, Boca Raton, FL.
10. Dook, J. 2006. Forensic investigations, blood spatter, properties of blood, teacher background information. http://www.clt.uwa.edu.au/_data/page/112508/fsb05.pdf (accessed June 25, 2010).
11. James, S. H., and J. J. Nordby. 2005. *Forensic science: An introduction to scientific and investigative techniques.* 2nd ed. CRC Press, Boca Raton, FL.
12. Gerhardt, P. M., and Gross, R. J. 1985. *Fundamentals of fluid mechanics.* Addison-Wesley, Reading, MA.
13. Larson, B. (Ed.). 2001–2010. *Surface energy (surface wetting capability).* NDT Education Resource Center. The Collaboration for NDT Education, Iowa State University. http://www.ndt-ed.org/EducationResources/CommunityCollege/PenetrantTest/ PTMaterials/surfaceenergy.htm.

Surface Pairings

7

7.1 The Value of a Surface Pairing

7.1.1 Hypothesis

In most forms of scientific analysis, a great deal of attention is paid to conditions. Both specimens and equipment are carefully scrutinized before and after use. The purpose of the attention is to isolate variables in observable (experimental) situations so that a particular statement of cause for an effect can be identified.

Experimental observations are usually further confirmed by additional testing and the use of control samples. In forensics, we are seldom presented with a pristine and identifiable set of ingredients, and only rarely is there time and resources available with which to experiment with reconstructions.

The idea of creating a hypothesis based on observations after an event has occurred is not new to science, but it is an integral part of daily forensic examinations. Many forensic conditions are not practical for experimentation. This problem can be solved, in part, by developing greater knowledge of how common materials and substances behave.

Conventional mention of paired surfaces conjures images of Teflon™ on steel (an example of very low friction), steel on steel, or rubber on dry pavement (an example of relatively high friction). It is well known that lubricants greatly reduce friction; further, water or blood can act as lubricants that are often common to crime scene evidence. Studies of these and other properties can be used to perform experiments at times removed from the urgency imposed by casework.

The forensic study of paired surfaces begins with definitions and concepts, many of which are not currently used in forensic literature. The precise language of choice is not as important as realizing that there is literature and knowledge, already in existence, that will assist in developing a more comprehensive expectation of how surfaces behave when paired. The knowledge begins with fundamental concepts of the forces involved and even how one should consider thinking about the materials.

The stresses and deformation that arise from contact between two surfaces is a field of study known as contact mechanics. Tribology is the study of friction, wear, and the effects of lubricants. The study of the behavior of liquids and soft solids, including viscosity and visceolastic behavior, are within the scope of rheology. All of these subjects are found under the umbrella of a fast-growing department of engineering known as material science or material physics.

As pointed out in Chapter 2, nanotechnology is contributing advances in the creation of ultrasmooth surfaces. In Chapter 6, we learned that the microscopic surface roughness of even smooth surfaces such as glass cause most fluids to form a uniform shape with a discernible contact angle. It is reasonable to assume that the contact angles for liquids will be affected by the smoother surfaces derived from nanotechnology.

Examples of microscopically smooth surfaces at crime scenes must be relatively rare now, but that observation has the capacity to change as new uses for nanosurfaces are developed. The study of common paired surfaces can yield a great deal of evidence.

7.1.2 Definitions and Concepts

In our study of surface pairings, a number of definitions and concepts need to be clearly defined. What follows is a conglomerate of terms and concepts derived primarily from the areas of study listed. This list of terms is intended as a frame of reference rather than a comprehensive guide.

- *Materials science or materials physics* is the scientific study of materials. The areas of study are concerned with understanding the characteristics and properties of materials, substances, and their behavior. The purpose of materials science is directed to the design of new materials or the improvement of existing materials.
- The *space* occupied by the bodies is a geometric relationship that defines the parameters of the contacting surfaces in terms of linear and angular measurements that create a frame of reference in space for that contact. The coordinates fit within Cartesian coordinates, and for the purpose of most forensic concerns, the measurements confirm applications of Newtonian mechanics.
- *Conforming surfaces* exhibit a more or less complete area of contact, as seen with two flat plates or a ball in a socket, whereas *nonconforming surfaces* tend to exhibit point source contact at one or perhaps a few points. These distinctions become important in the interpretation of wet impressions.
- *Time* is the measure of a succession of events of some chronological description that can be considered an absolute quantity for which there are sufficient details to permit such definition.

- *Mass* is the quantitative measure of inertia or resistance that can be, at times, expected when conditions allow the calculations to be precise, allowing some evidentially useful determinations.
- *Force* is the vector action of one body compared to another. Typical forces include those that are tensile, compressive, tangential, electrostatic, or are derived from impact, friction, and heat. Properties generated by the application of some force reveal such characteristics as thermal conductivity or expansion or electrical conductivity or resistivity, for instance.
- *A normal* describes the direction of a force as essentially at right angles to the plane of the surface against which it is applied (whether the contour of that plane is flat or curved).
- *Vector* and *scalar* quantities are directly related in that a vector has a scalar magnitude such that $\Delta V_2 - V_1$ would be representative of (Δ), the amount of change, between two vectors. That calculation would in turn produce a different but proportionate change in relative scalar magnitudes shown as $\Delta V_2 - V_1$.
- *Friction* is the resistance encountered by a body (or particle) while in motion or as a force begins to act on that object. Friction is then a measure taken at the interface between surfaces. Friction between two solids is one of the most common concerns to forensics, but calculations can be further influenced by the presence of liquids, gases, or combinations of factors.
- The *coefficient of friction* is a measurement of either static or dynamic friction for a particular surface in contact with one or more other surfaces; it is typically measured by a weight ratio apparatus, spring balance, incline plane, clamping, pendulum, or motorized tribometer. Undergraduate engineering students are commonly advised that a given coefficient of friction for a particular material is a starting point in assessing the actual forces involved in a specific equation, and further that the coefficient of friction for metal is not applicable to polymers.
- *Deformation* occurs in three primary forms that apply to forensics; all are greatly affected by the presence or absence of heat. *Elastic deformation* is the temporary distortion of a material that exhibits the ability to return to its original shape on removal of the force that caused the deformation. *Plastic deformation* is the resultant exhibition of permanent *strain* in response to *stress* (an applied force). *Creep* is a separately permanent form of plastic deformation that occurs over a relatively long exposure period to the force that is causing the deformation to occur. *Deflection* is the magnitude of bending in response to a given normal. While these terms all apply to both metals and polymers, the calculations and forces involved

are different according to the properties of the materials involved. *Contusions* or *bruising* are hematomas (a condition in which blood is observed in tissue outside the vessels). These markings appear as generally diffuse deformations relating to the normal of an applied force (sufficient to injure), whereas *lividity* (observed in cadavers) is a similar and more temporary condition (that occurs closer to the surface of the skin) and often retains some comparatively detailed records of prolonged contact (caused by the blood being restricted by pressure from pooling in the contacted tissue). Plastic deformations could almost extend to scars in a biological specimen.

- *Relative motion* includes *sliding* (the linear velocity between surfaces), *rolling* (the angular velocity of two bodies around a common axis), and *spin* (angular velocity about a common normal). Motion in space further makes use of the longitudinal axis of a single object traveling in a given direction, which provides definitions of *roll* (rotational movement along an axis), *pitch* (the front-to-back tilting of that same axis), and *yaw* (which is the skewed or sideways orientation of the axis relative to a direction of motion). The first three terms refer to relative motion in which both objects are in motion. The last three terms tend to refer to when the motion of a single body or object is considered as if on a stationary, frictionless plane.
- *Material strength* is measured in the mechanical properties of a given material to withstand the application of forces on it. *Hysteresis* is the slow return of a material to its original shape or form (up to the yield strength) once a *load* (acting force) is removed. The *yield strength* is that point at which (if exceeded) the same material can no longer exhibit elastic behavior with the removal of the load. *Compressive strength* is the ability to resist ductile or brittle failure in response to compressive stress. *Tensile strength* is the strength of a material in resistance of a tension that tends to pull on that material, thus resisting elongation. *Impact strength* is essentially the ability of a material not to fracture or rupture in response to impact. *Fatigue strength* is the strength of a material to sustain cyclic loading without deformation, as opposed to the continuous stress that causes creep. Tensile, impact, and fatigue strength are also measures of the *toughness* and *hardness* of a material. It is interesting to note that there are parallels between the molecular weight of a polymer and its strength. The effect is that a material of higher molecular weight will tend to exhibit more secondary bonding between the polymer chains, thus making the material more resistant to deformations such as creep.
- *Forensic engineering* is the investigation of materials, products, or structures for which failure can be related to a material. These matters are normally beyond the means of most practitioners to investigate

since they involve the use of specific microscopic or metallurgic knowledge and evaluative techniques, but as equipment is obtained, such as scanning electron microscopes (SEMs), those in the forensic sciences can expect to experience more use of these increasingly common technologies.

7.1.3 Properties of Common Substrates

Material systems are routinely designed to suit a purpose. Forensic practitioners are familiar with the range of engineering dedicated to the design of an automobile tire or the design of a footwear outsole. Many other surfaces have also been engineered so their properties meet certain design criteria. We have seen some of this engineering at work with the design of agglomerated materials such as concrete.

Nature has presented us a wide range of surfaces and textures, some of which invite mimicking with engineered products and all of which have a role to play in contact between two surfaces. A few other methods of engineering materials seem worthy of mention due to their role in extending or altering the way in which a substrate behaves. These include the introduction of reinforcing materials, surface modifications such as compression, surface coatings, and chemical and thermal treatments.

Thermal treatments, such as annealing, are common for metals that are intended for further machining. This treatment works well because the material retains its relative softness and some flexibility. Other processes are used to introduce hardness and greater strength for specific purposes.[1]

Mechanically modified surfaces have usually been subjected (as with a ductile material, such as metal, that can be drawn into a wire) to compressive forces. The effect of compression on a ductile material is reduction in the susceptibility of the material to fatigue. Compression is also one of the factors in the production of tempered glass.

Traditional surface coatings may be metallic, polymeric, or ceramic and are chosen for design purposes that make use of the protective or decorative properties of the coating. The selection of an appropriate combination of substrate and coating has much to do with the formation of useful bonds between the substrate and the coating or a mechanical attachment to the roughness of the surface of the substrate.

7.1.4 Properties of the Common Matrix

If we accept the common forensic usage of the word *matrix* as referring to a substance intervening between two surfaces, then we must consider the most common matrix to be water. Water is present, to some degree, on all surfaces within our atmosphere, with the possible exception of a vacuum. In many of those situations, the water takes the form of a thin adhered film, not much more than a molecule thick.

The presence of thin films of water is, for many purposes, ignored. The effects of adhered water vary from one surface or condition to another. If it were possible to ignore adhered water, water plays a role in many other scenarios. Even disregarding the familiar changes of state, water also performs as a lubricant in some instances and a solvent in others. As the singular most relevant matrix, water bears more scrutiny.

The molecular composition is worth noting since the slightly higher negative charge of the single oxygen atom leaves this covalent compound a polar molecule with a dipole moment. The interactions of dipoles on each end of the molecule account for the presence of a skin (surface tension), and the net charge further explains a predisposition to adhering to other materials (hence the term *adhesion*). The capillary action of water, vital to plants, and the designation of water as "the universal solvent" also enhance the status of water as a matrix.

Water is considered an unusual substance because most liquids that solidify do not float once immersed in a liquid. Water (with the exception of somewhat different values and characteristics for saltwater) exhibits a maximum density at about 4°C. This means that the density decreases as the water is either warmed or cooled (the change in density in solid ice is explained by the expansion of the substance, which can range up to about 9%).

Saltwater freezes at about –2°C. As the temperature decreases to the freezing point, the density of saltwater increases. Exact figures are attached to specific samples since the degree of salinity varies according to the source. Salinity has no great effect in most forensic cases but should be tested when it could cause a shift in useful values. Brine can generally be simulated in testing with an approximate 3.5% (average) to 5% (relatively high) solution of salt.[2]

The boiling point of water varies with barometric pressure, which has some relevance to properties in different locations (above or even below sea level). In gaseous form (water vapor), water can routinely affect the development of evidence (with ninhydrin analogues or cyanoacrylate treatments in particular) and should be noted as a minimum and maximum value in laboratory validation processes.

The second most common matrix cannot be named as anything other than a set of variables labeled here as contaminants. Rarely, sometimes due to obvious associations, a contaminant can be predicted to be a certain substance, as with evidence of a suspect's greasy-looking fingerprints subsequent to consumption of a bag of chips. This detection can apply to footwear or other markings that are subsequently consistent with a likely exposure to some readily available contaminant.

Lacking obvious clues, we are left to examine and draw conclusions from the appearance of the marking. The comparison must begin with some relative experience to the appearance of markings and the behavior of substances under load or stress.

7.1.5 Synergistic Effects

For health and safety reasons, it is worth instructing that combination of water with electropositive elements (group 1A of the periodic table), including lithium (Li), sodium (Na), potassium (K), and cesium (Cs), will result in the release of hydrogen (H), which is a flammable gas that represents a potential risk for either combustion or explosion hazards. Further, a strong hydroxide is a caustic material posing a risk of burns.

The training of new practitioners could benefit from some familiarity with the appearance of the potentially reactive elements mentioned and with their compounds. Identification of the likely presence of those elements in the various types of scenes common to the workplace (geographic working area) could prevent unnecessary exposures and is as worthwhile (as a concern for health and safety committees) as ensuring that bleach is not used in conjunction with cleaning fluids that contain ammonia.

7.1.6 Expectations (and Distortions)

The concept that footwear can be identified from striation markings is relatively new. The possibility has been noticed by a few and used by even fewer. On the face of it, this does not seem a momentous discovery, but it does represent an opportunity to utilize what was previously considered simply "distortion."

A marking that is distorted or contains elements of distortion needs to meet the criterion of any form of evidence: It must possess measurable and reliable detail that can be ascribed to a source by the use of corresponding features. The concept that a blanket term, *slippage*, could be assigned to a particular area of friction skin impression would raise the hackles of many seasoned practitioners, who would expect a better explanation of what is distorted, how it appears, and whether it is deemed to interfere with conclusions. Statements containing anything less are derived from a lack of knowledge or a lack of experience.

Parallel striated markings are used on a day-to-day basis. The striations referred to are the bar codes that allow merchants to identify a product and its price and keep track of inventory. There are practical and ethical reasons that govern the widespread use of friction ridge impressions in a like manner.

Striated footwear has been described, as used in casework, based on no more than an estimated similarity between the found three-dimensional markings compared to a hand-generated test impression. The test impression was obtained by rolling a suspect's shoe outsole in modeling clay. That is junk science, not because it could not work, but rather because it has not been proven or tested, and the resulting comparison is not quantifiable.

The credibility of the evidence further requires some assurance that the exhibit (suspect's shoe) has not been altered by the process (thereby eliminating any chance of reproducing the effect as described) or that the results are not compared to biased test methods. Technology may solve this problem by allowing striations to be further explored without direct physical contact. Computer extrapolations that accurately generate a replication of any chosen edge in striated form could provide an unobtrusive comparison technique.

The current path of recording friction ridge details is elementally flawed. AFIS (Automated Fingerprint Identification System) technology is designed to treat friction ridges as two-dimensional entities on a glass substrate, which mimics many taped lifts and chemically developed impressions under "normal" circumstances. This methodology limits the accuracy of the recorded images to these more or less "ideal" situations. Three-dimensional scans that can be attenuated to produce two-dimensional images could, one day, provide information regarding the angle and amount of pressure applied with any impression.

Testimony that contains reference to the way in which an exhibit was held or touched adds a new dimension to the evidence. Ridges contain structures known as ridge units. The asperities of multiple ridge units are capable of forming striations, and the use of this "distortion" is limited by technology and imagination. The future use of laser scan point clouds may change this.

7.1.7 Another Challenge

Discovery often leads to more discoveries, and in that tradition let us consider that friction ridges offer topography not unlike that of a worn footwear outsole in respect to the generation of markings. Friction skin subjected to pressure, as in the markings left by someone attempting to force open a sliding glass window, leave a type of distortion commonly known as slippage. These striated friction skin markings (with no ridge detail) are often found in slippage.

If this feat were achievable, it would most likely begin with a palm imprint for which one can see the shape of a palm and the associated digits and thus have the advantage of orientation for the segment of the palm that caused the marking. Striated palm prints are perhaps the most common of such markings; unfortunately, palm prints are not yet included in friction ridge databanks. Computer applications would be required to generate a topographical slice of the suspected area of contact, much as can be achieved now using a portion of a point cloud image to isolate just one segment of a given topography.

It is the inherent flexibility of friction ridges or footwear outsoles that complicate location of a source. Edge characteristics of footwear may occur at any point along the width or length of the edge or its adjoining surfaces. This aspect could be applied to the asperities of ridge units found in friction skin; the tips of the last ridge units in contact would form the shape of the marking. Locating the precise combination and orientation of ridge units that could be predicted as donating features would be a highly difficult task.

Identification based on striations would represent the first use of ridge details based entirely on the occurrence of alternating asperities and voids. The likely method of comparison would be to match computer-generated striation patterns to crime scene impressions. It is currently possible to create a cross section of any named portion of a point cloud image. It is further possible to proportionately manipulate the cross section to conform to the shape of a given substrate.

This is not technologically possible now. One day, that prospect may not seem so distant. Remember that it was not so long ago that searching for a fingerprint using a computer was the stuff of fiction.

7.2 Sequential Analysis

7.2.1 Understanding the Sequence

Many techniques that work well under one condition will not function as well under another. Destructive aspects of techniques and unknown conditions of an exhibit require that treatment be conducted in a sequence that optimizes chances of success in the development phase of examinations. The evidence, as it is found, can also be the product of sequential circumstances.

The sequence of events can be determined with an impressive degree of certainty where one event is disturbed by another. This sort of determination is a matter of observation that merely indicates the order in which one event is preceded by another (see Figure 7.1).

Lawyers and judges respect and increasingly demand DNA evidence for each case. This preoccupation may contribute to a devaluation of other forms of evidence. This amounts to a form of bias.

A DNA specimen that holds considerably less evidential value is often favored over a form of directly associated impression evidence. One can only hope that this type of abuse of evidence will subside as the technology gains wider understanding.

Imagine the future of DNA technology as an ever-decreasing amount of specimen required to extract a useful genome. One day, in the not too distant future, it will be possible to extract DNA from fingerprints. The sequence of one marking over another suggests that the dominant impression was made last. DNA evidence does not discriminate this form of sequence and may offer markers that associate much older and unrelated events.

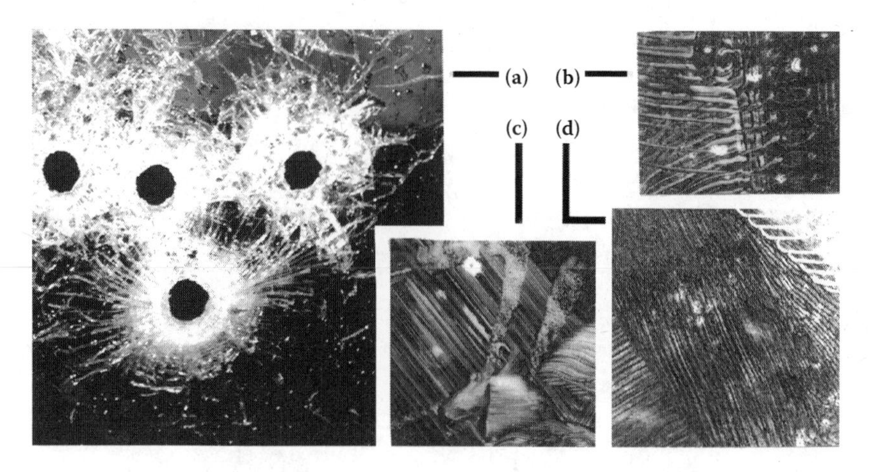

Figure 7.1 Sequential events. Four examples of situations in which a sequence of events could be inferred by the state of the marking. (a) A sequence of bullet holes. Careful tracing of the fracture lines can allow discrimination of the sequence in which these holes were made. (b) Striations leading to and away from an impression of a watchband. (c) Striations caused by footwear that are interrupted by a direct contact of pattern elements. (d) Striations leading to and away from a friction ridge impression. An added artifact at the top right of this specimen was caused by movement of the teeth of a comb that obviously removed and thus interfered with part of the friction ridges.

Proponents would extol the virtues of the refined process, but practitioners with more experience would advise caution. It is unknown how many of the fragile latent markings found or missed do not include a DNA component. It is further likely that of those that could possess some alleles, many more markings will lack a sufficiently abundant source.

An article regarding the uses and vulnerabilities of cyanoacrylate fuming techniques expressed a legitimate concern regarding the blanket acceptance of a sequenced technique. Referring to barely visible latent markings[3]:

> Sometimes the latent can be developed by another technique, but on other occasions it is lost forever. This was the case with a piece of foil that had been used to package a suspected drug substance. A visual examination of the piece of foil revealed indications that a good latent print was present. CA [cyanoacrylate] comes well recommended as a technique for foil and so, into the tank went the piece of evidence. A later examination of the foil did not reveal any latent print being developed by the CA fumes. (p. 5)

This account, written over a quarter of a century ago, was in no way intended to belittle the attributes of cyanoacrylate fuming as a technique. The intent of that article was to point out the vulnerability of placing all of one's eggs in a single basket. DNA or any other form of evidence is no less susceptible to misinterpretation without the support of corroborative evidence.

Also, on the same page as the previous quotation, there is mention of a case in which a kerosene lamp was found contaminated with "an oily substance." The exhibit did not respond to laser examination and was subsequently treated with cyanoacrylate. Useful markings were located, and it was further observed that the treatment had apparently "dried up" the fluid. The observation of the effect of glue fuming on the "oily substance" is most interesting.

A practitioner with reasonable experience can tell you that friction ridge impressions or even footwear impressions are unlikely to survive the destructive effects of a typical car wash. That it is possible for a marking to survive at least some inhospitable events does not mean that such an event would not be accompanied by at least some obvious deformation (Figure 7.2).

Figure 7.2 Car wash experiment. The overall image shows the test vehicle prior to entering the car wash. Inset shows an enlarged version of the only marks to survive exposure. They are placed to the left of, and in the same orientation as the original footwear and thumb on the hood. The thick coat of petroleum jelly has been reduced to a thin film by the wash.

In an attempt to illustrate the vulnerability of some markings, a simple and informal experiment was devised to test the durability of impressions exposed to commercial automatic car washing. Three different automated carwash establishments were chosen. The same vehicle was used for each exposure. Footwear and palm prints were created (they were located within boundaries drawn alternately with a china marker, lipstick, and crayon) on each hood, fender, and windshield of a vehicle and were reapplied to almost exactly the same locations between each session.

There is a single change to the procedure that is pictured in Figure 7.2. This change involved the structure of the car wash; the last of the three facilities did not use any brushes, just high-pressure spray. This was the only situation for which any amount of evidence survived. It was noted that the slightest physical contact caused removal of the surviving marks in the contacted area.

These results reinforce the concept that a typically created finger or footwear imprint is a fragile piece of evidence that would require some drastic alteration to survive the hostile environment of a mechanical car wash. Believing that an average fingerprint is capable of surviving unscathed where other more stable substances are obliterated would not be a reasonable conclusion.

7.2.2 Effects of Pressure on Sequential Analysis

Every marking is created in the presence of some form of pressure. Physics and chemistry recognize that atmospheric pressure can have an effect on experiments; further, small forces are often discounted. Large forces can also be overlooked.

The amount of pressure exerted by a locomotive engine is calculated by dividing the bulk weight by the number of wheels and some determination regarding the amount of area for each wheel in actual contact with a rail. Typical problems with these calculations include variances in topography from one wheel to another. Problems aside, a general estimate of pressure was derived as approximately somewhat less than 100,000 kPa pressure exerted by each of the eight wheels of an engine known to weigh approximately 103,040 kg. In the following experiment (see Figure 7.3), this locomotive was used as a tool to compress a series of coins interspersed with other substances.

7.2.3 Equipment

1. A locomotive on rails.
2. A series of coins in the following denominations: penny, nickel, dime, quarter, $1 coin, and $2 coin.

Figure 7.3 Locomotive and wheels. A locomotive was used as the tool in a simple experiment in which the effects of pressure and the transfer of materials under pressure were considered.

3. Viscous substances included axle grease, blood, motor oil, and modeling clay.
4. Measuring and recording devices (measures, chalk, and a camera).

7.2.4 Methodology

The wheel was measured and marked with chalk, and a test area was selected that represented slightly more than 1.5 revolutions of the wheel. The test specimens (coins weighed in advance with an analytical balance) were arranged on the track and photographed prior to subjecting them to compression at speeds not in excess of 5 km per hour. The results of exposure to contaminants (the specimens) were examined and photographed following the completion of one full rotation of the selected wheel.

7.2.5 Discussion

The intent of this experiment was to examine the effects created by a great amount of pressure. Some materials were expected to transfer, but the amount and method of transfer was unknown. Obvious transfer was seen in the experiment to cause a secondary impression at one complete rotation of the wheel (see Figure 7.4). The coins were impregnated with foreign materials that skewed the resulting weight of the coins. Some material was observed to have been transferred by at least some of the coins, but the cross contamination did not allow accurate calculation of the amount of material transferred.

It was further noted that viscous materials were visibly transferred over the rotation, and while blood and oil were not visible, it was considered

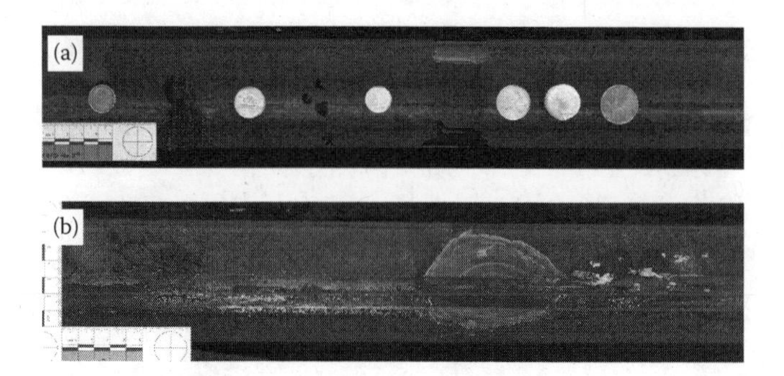

Figure 7.4 Test materials on rail. The arrangement of the test materials (a) prior to contact and (b) postcontact transfer of that material following one complete revolution of the wheel.

reasonable to assume that further testing would prove the existence of a trace amount of those fluids. Further testing would be required to determine the longevity of a sequence of transfers over a distance of rail or the effects of increased velocity.

This experiment displayed the effects of an amount of pressure on materials seldom experienced at scenes of crime. The sequence of the materials was somewhat blurred in the result. The experiment was conducted twice since in the first instance the locomotive driver had applied brakes a little prematurely. The braking action on a rolling train wheel introduces an amount of frictional variance known as *creep*.

7.2.6 Creep

In the experiment shown in Figure 7.4, a different topic was illustrated by the wheel in use. Creep is the elastic response between two materials experiencing tangential strain. The effect, seen in the illustration, is an alternating series of stick and slip (Johnson 2003).[4]

The phenomenon was first noted by Reynolds (1875).[5] In his experiments, he noticed that the circumference of a wheel measured consistently short of the distance traveled in one complete rotation.

Creep was recorded in the previous experiments (resulting in Figure 7.4). The measurement of the distance traveled was substantially longer than the calculated diameter of the wheel. These measurements were intended to predict the location of a transfer point on the rail following a complete rotation. The discrepancy was at first considered a simple error and nearly disregarded. The prediction was important to the methodology; the wheel could not easily be viewed from the cab of the engine and measurements were intended to create a chalked landmark on the rails by which to judge the necessary distance to slowly move the locomotive. Chalk markings set on the wheel and on the rail repeatedly confirmed the presence of some issue that was later explained by finding the definition of creep.

7.3 Transfer

7.3.1 Direct and Anecdotal Transfers

Transfer of one substance from one location to another requires contact, which is further defined by some degree of force and pressure. A consequence of such conditions may cause a direct transfer that may result in either removal or deposit of materials between substances. A direct transfer of materials can be thought of as any situation that results in the creation of either visible or latent impressions.

Examples of direct transfer include the most common of impressions, such as fingerprint or footwear impressions. There are few references to the amount of materials involved in such transfers. Measuring a fingerprint is a somewhat difficult task that tends to explain the fragile and delicate nature of impressions.

Students in science classes are routinely cautioned against leaving fingerprints on glassware, citing that the added weight of fingerprints could skew the results of subsequent measurements. This theory was put to the test using an Ohaus, Analytical Plus, model AP II CS, accurate to ±0.0001 g; a pair of clean tweezers; and a clean glass microscope stage. Two separate hygrometers confirmed an ambient temperature of 23.3°C and 63% relative humidity.

Repeated tests involving the placement of a single visible fingerprint created by rubbing a thumb repeatedly through hair at the nape of the subject's neck did not result in any reliable measurement. Weighing of two similarly visible fingerprints with a high lipid content yielded a reading of 0.0001 g. Several subsequent impressions resulted in a reading of 0.0004 g.

The results demonstrated that if glassware is mistreated, the lack of cleanliness might have a deleterious effect on the accuracy of measurements. The incidental observation that a typical fingerprint would not likely be as heavy as the enhanced versions was proven by attempting to weigh numerous overlapped markings without success.

Indirect transfer could be represented by the family of markings also known as voids. These marks are indirect in that some material has been removed from a surface in a manner that either possesses attributes of detail or at least indicates the likelihood of recent contact (as in voids found within a spatter pattern that indicate the presence of an object that was moved after the event). Dust imprints are sequentially opposite in that the void is removed after the coating of dust was created.

7.3.2 Deformed Transfer

Deformations are a normal and often temporary product of contact. The nature of the impression that results from contact will offer clues warning of the presence of deformation only in the most extreme cases. When the overall effect is minimal, or contained to only a nonrelevant portion of an exhibit, there may never be a reason to refer to such small differences.

While explainable differences will not have an impact on testimony, more complicated interpretations would benefit from understanding the methods by which deformations can have an impact on the appearance of an impression. Let us take, for example, a crime scene footwear imprint that comprises a small portion of the patterned area and exhibits considerable disparity in measured size with test impressions of the suspect's shoes.

One conceivable explanation for the differences may involve transference of deformation. While lifts or photographs of the marking may not convey the cause of stress, a review of casts or the scene itself may show that for the suspect's outsole to create a similar marking, sufficient stress would be required to create deformation. These conditions are definable, and any interpretation of apparent dissimilarities would require some attention to creating test impressions that accurately reflect the presence of stressors.

7.3.3 Schallamach Patterning

The fine and often wavy patterns that are occasionally observed on the surface of rubber products can originate as a product of stress on the material. A collision between two metal surfaces, as is too often the case with motor vehicles, results in visible damage to both surfaces. Interactions between a hard and a relatively soft surface under the same conditions produce different results. When the conditions of contact are abrasive, the softer material can be subject to microscopic shear, thus creating the wave patterns known as Schallamach waves.

The condition is of a temporary nature when similar stressors are encountered. The patterns shift not only in details but also, at times, in direction as a response to the angle at which the frictional force is encountered. These waves are associated with both footwear outsoles and tire surfaces.

Control of friction in a synthetic material is a commercially relevant area of study that has received much attention from researchers. There are some basic principles that have become accepted in these studies, such as the recognition that elastomeric friction can be described by both tribological (interfacial) and rheological (bulk material) models. There is no clear indication regarding which set of influences is the dominant cause of the waves.

Interfacial studies reveal the role of formation and breakage of adhesive molecular bonds. Glass is one surface that exhibits microscopic roughness (with tops that are known as asperities) that defines the amount of friction between interfacial surfaces. The amount of friction can also be affected by the softness (lower modulus) of a rubber product, which would logically develop higher friction due to contact with greater surface area. Experimental observations contradict that expectation.

The role of bulk materials is explained by consideration of the internal friction between molecules of an elastomer under periods of stress and relaxation. The polymer chains of an elastomeric substance will give off heat on applications of stress (observable by stretching a rubber band), and the resistance of chains and molecules may also contribute to the slowed return of rubber to its original shape.

What seems well accepted are the following:

1. As temperature increases friction decreases.
2. An increase of sliding velocity will generate an increase of frictional force.

Schallamach waves have been described as consisting of "tunnels" of air that disrupt contact (which sounds similar to the hydroplaning of tires) "rather than the instantaneous interfacial failure found with stick-slip."[6]

It is possible that a particular mold could have been fabricated (made) using an affected surface as a model for casting the mold, or that the product was affected by friction on release from the mold or a set of rollers. Such markings would represent a transfer of detail from a specimen that bore the original patterning. This also would explain the appearance on new shoes of Schallamach patterns subsequently erased by normal wear.

7.3.4 Intaglio

The art of printmaking offers an opportunity to examine methodically the role of liquid matrix on a substrate. Many of the techniques and materials date to antiquity. The variations known as the intaglio process are typified by etching a matrix or plate, which is then inked and wiped before a final "pull" is made by which the ink is pulled from the etched areas of the plate by the receiving surface.

The experience gained from printmaking clarifies the mechanisms by which an impression behaves. No matter whether the source is a finger, outsole, tire, or tool, experimentation with printmaking creates easily repeatable examples that model the states in which impressions appear. A tonally reversed marking could reasonably be termed an intaglio process in which the furrows of friction ridges or the voids of a patterned surface supply the detail.

Lessons learned from printmaking assist in understanding more than tonally reversed markings. An impression composed of mixed tonality (one of the most challenging types of marking to analyze) is much easier to interpret with knowledge derived from printmaking.

7.4 Coincidental Impressions

Coincidental impressions such as a palm print and one or more corresponding fingerprints are often found in a position that seems consistent with having been made at the same time. It can be more defensible, given the larger area, to claim that the elements were created simultaneously as compared to two or more fingerprints, commonly referred to as a "cluster." This distancing of the elements can present an ethical dilemma for the examiner.

Formation theories may begin with taking measurements of relative locations and orientations of the impressions. These may logically indicate that the imprints were made in a contemporaneous fashion. The identity of the source may be evident in one of the elements.

Where each element is linked by sufficient and continuous detail along the phalanges (for instance), there should be no difficulty in associating those partial fingerprints with the palm. The use of supportive measurements should not be confused with a method of certainty in associating single elements to a supposedly greater whole. Measurement, observation, and accounts of events may combine to form an image of how the acts were committed. Damage, dimensional stability, and other distorting features must be scrutinized and explained as part and parcel of the formation theory of any impression.

In short, observations must be supported by either data or experimentation. Theories should be scrutinized and proof found before one is required to give evidence. Verification of the findings and revisiting the case may prevent the embarrassment of error.

Coincidental impressions must be treated as separate events without the benefit of additional proof. This principle must be applied to any form of impression evidence. One should generally attempt to confirm that there are no clues regarding the conditions under which the markings were created that contravene the formation theories.

Impressions of an apparently coincidental nature containing or affected by a void may, in some cases, be associated by a physical match between the portions of an impression and an intervening object that entirely or partially completes the affected area (see Figure 7.5).

Multiple contacts between surfaces are routine in any crime scene. Where the details permit, a series of contacts may further reveal evidence of pressure, distortion, the presence of contaminants, indications of a timeline, a sequence of events, or alternatively, obscure the usefulness of one or more contacts or details. This is discussed in further detail in the chapter dealing

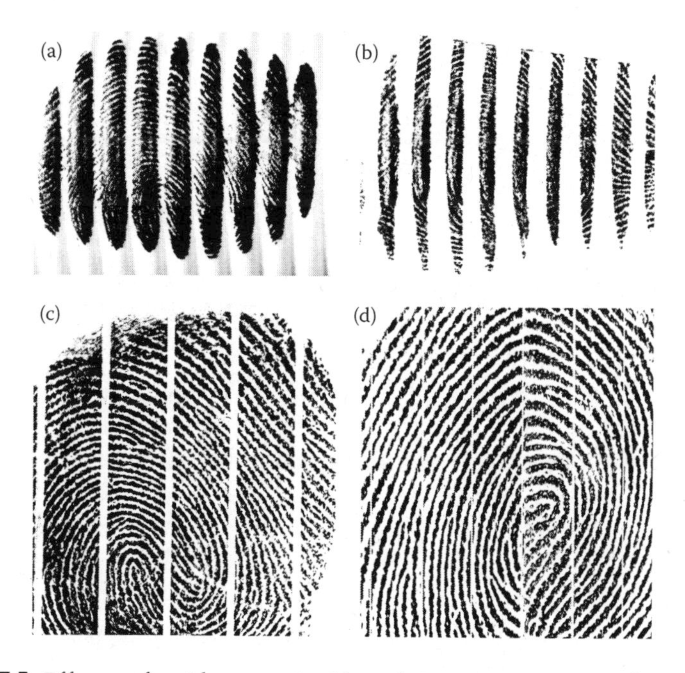

Figure 7.5 Effects of voids on coincidental impressions. Four friction ridge impressions are shown; all were made with the same digit, and all are discontinuous to varying degrees. These markings pose a range of obstruction to identification. Two of these markings are considered unsuitable for comparison as presented. Impressions (a) and (c) possess sufficient correspondence of clear details that do not support a theory of having been made by any other source. The two markings on the right are another matter. Impression (b) has good but widely separated detail that would be difficult to associate; the version in (d) is actually two impressions made by the same digit and would need to be resolved before a comparison could be made.

with distortion. This chapter is devoted to the features of a single event for which the surfaces are paired at the moment of a particular contact.

Exposures, however short, that occur between two materials can have predictable results. The ability to form some conclusion about the product of a contact requires knowledge of the expected properties of those combined materials in any situation and the specific condition of the surfaces in the incident under consideration. A database for commonly encountered materials, both in the expected form and with reference to some known aspects of degradation, would be vital information in the analysis of a single contact. It is outside the scope of this book to provide such a database, but it is possible to examine how some basic materials behave.

Enhancement of impressions has been and will remain an ongoing study. Techniques are most often developed to improve details of a specific substrate under a finite set of conditions. Events that involve multiple interfaces must first be recognized as such and often treated as separate pairings. The choice of techniques may be influenced by whether the impression is suspected of resulting in adsorption, which results in wet deposits or dried stains on the surface, or in absorption, which is seen with wet deposits that have migrated into a porous substrate such as paper. Remember that both can occur in a single contact event.

7.5 Paired Surfaces at a Point of Contact

The occurrence of two surfaces coming into contact is most often a singular event. The interactions between two surfaces at a point of contact may be thought of as a pairing of those two surfaces for whatever duration the contact lasts. If one of the two materials straddles the second material to effect contact with a third or subsequent surfaces, those contacts are each pairings of two surfaces. Even though there may have been just one motion or action, there could well be simultaneous pairings at other points of contact, resulting in multiple pairings from a single incident.

Isolation of paired surfaces is extremely relevant to training and the presentation of evidence. Seasoned examiners will have little trouble in applying this separation to those impressions that may require divergent treatment appropriate to the materials involved to enhance the evidence (see Figure 7.6).

Rudimentary pairing studies also afford a chance to analyze the mechanisms of the contact. The study should begin with the natural state of the substrate as it would have been found, the changes imparted by the type and duration of contact, and the observed results of contact. This type of study may involve physics, biology, chemistry, or engineering in any combination (see Figure 7.7).

7.6 Industrial Materials

7.6.1 Common Attributes of Manufactured Products

Typical industrial materials are families of commercially viable substances that meet specific needs by design. While wood is a natural material, lumber is not, and the defining point is the application of manufacturing, which alters the raw material into a product (not only by shaping and drying processes but also by chemical treatments). The families of industrial solids include fabrics, wood products, metals, rubber, and polymers, to mention a few. Industrial liquids further include solvents, detergents, oils, and waxes.

Industrial materials interact at given temperatures and pressures with predictable results. The presence of a particular contaminant, such as

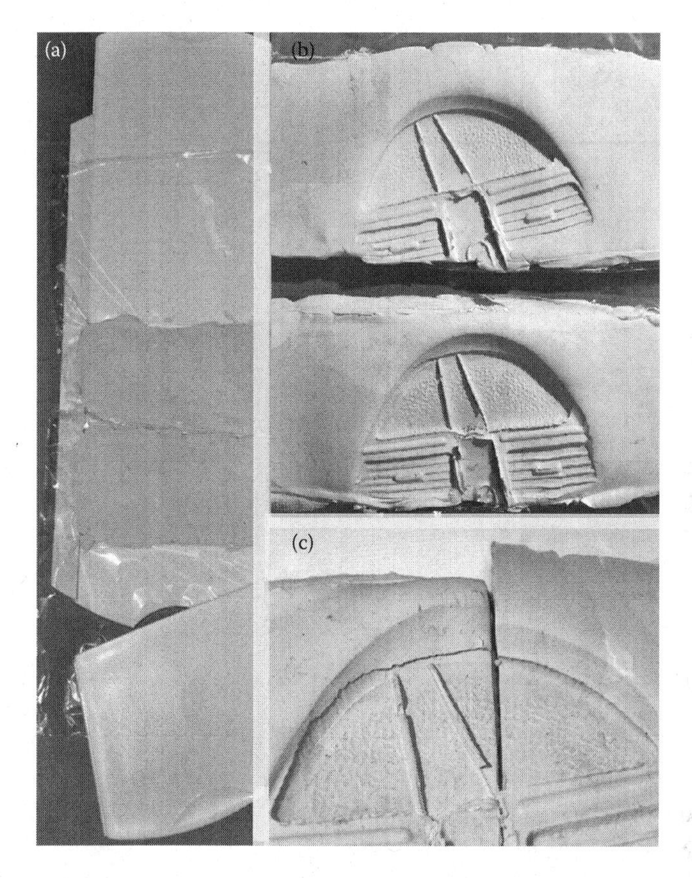

Figure 7.6 Illustrated transfer of force in clay. Clay was shaped in a frame (a). The effects of a heel imprint without confinement (top of b) and the same pressure applied in the image (bottom of b) with the clay bounded by the molding frame. The resulting impressions were cut and juxtaposed for comparison (c). Note the creation of an up-thrusting feature where the material is bound in other directions.

adsorbed water, may cause an effect that certainly is not necessarily visible. Interpretation of how industrial materials affect impression evidence is a study of surface and interfacial processes.

> The chemistry and physics at surfaces and interfaces govern a wide variety of technologically significant processes. Chemical reactions for the production of low-molecular-weight hydrocarbons in oil are catalyzed at acidic oxide materials. Surface and interfacial chemistry are also relevant to adhesion, corrosion control, tribology (friction and wear), microelectronics, and biocompatible materials.[7]

Aside from clearly observable phenomena, one must resort to technological solutions to probe a relevant surface. Some common techniques, such as

Figure 7.7 Tractor tire on pavement. Transfer of pressure effects experienced by a tractor tire on pavement. In a softer substrate, the tire will have little reason to flex; on pavement, the stress causes substantial deformation of the sidewall.

optical spectroscopy (visible and infrared), may allow determinations such as the chemical and molecular composition of the substances involved. Surface analysis of a topographical nature is achieved through the use of an SEM.

7.6.2 Interactions at Interfaces of Industrial Materials

Industrial materials, produced in a variety of substances, are often patterned for either design or functional reasons. Interactions with a substrate may often result in a transfer of the pattern from the donor but that is not true in every case since these materials may cause or be acted upon by the substrate, thus sustaining features transferred from the substrate.

It is also common for a substance such as rubber to possess the characteristics of a hard solid (with softer materials) and yet display a different response under increased stress (as seen when paired with harder materials). This represents a large family of substances that would create two-dimensional impressions in one circumstance and three-dimensional impressions in another.

In the case of markings made on a flat surface, it must be realized that what is flat at the time of contact may be affected postcontact. For example, some materials, which are otherwise flat, will experience variable shrinkage when heated. With some of the more vulnerable plastics, this can result in a piece of sheeting that appears crinkled when left in a vehicle window for a few hours, is exposed to solvents, or receives an impact by some force (either quickly or over a period of time). Pressure applied to many surfaces, including foil, papers, and some plastics, may result in embossing with varying degrees of permanence.

There is deceptively little in the way of permanence in the world around us. Metal, a hard and relatively stable substance, will bend, stretch elastically, or expand under stress. Many other substances that seem stable are easily affected by specific forces.

With our senses, we are able to detect only a few of the most obvious examples of dimensional change. The precise boundaries of stability for various materials are defined by the reactivity of the substance under stress. Changes of state, malleability, and solubility are visibly detectable, while stretching, shrinkage, and swelling can occur in a subtle manner, often going unnoticed.

7.6.3 Causes of Change to Substances

Substances can change in response to stressors or the immediate environment. That which was soft can harden, brittle materials may soften, and the effect can be unexpected. Digging in a wet garden will provide examples, as the shovel will easily penetrate well-aged and water-softened specimens of wood or bone. Plastic substrate (such as soft clay) can be affected by such a light touch that, without directional lighting, the mark appears latent. The same clay with more or less moisture will behave quite differently. Some plastic substrates are highly elastic. Few people would expect to find an identifiable marking on a powdered donut and even less would expect a fresh example that is well-impressed into the surface to heal itself within hours, as the structure of the donut returns to the original shape and the evidence can effectively disappear. Chemical and physical changes are actually commonplace in forensic case work.

An example of impressions that can demonstrate the extremes of substrate change are markings that are left on "tacky" paint or varnish. These impressions are latent and respond to some powder development but are seldom connected to the commission of an offense. Unlike an etched marking, a chemically induced event, these imprints are plastic (three dimensional), and even though they could easily be misinterpreted as having been created more recently, they are usually well aged.

Simple differences between an etched marking and a plastic one are observable. Etched markings rarely respond to powder enhancement; they are only visible under specific angles of transmitted or reflected light, and when visualized, they correspond to direct contact with the ridges of the donor topography. Plastic imprints tend to have penetrated deeper into the substrate than etched versions, thus appearing visible from a wider range of angles. In some cases, the exposed apparent ridges (corresponding to the furrows of the source) can be enhanced with powder treatment or lighting techniques.

Footwear outsoles, tire treads, and wood products share a number of characteristics. They contain carbon and are formed as polymers. While the cellulose of wood is not to be confused with the output of polymer industries, there are useful analogies in the way these substances behave.

Most common forces that cause change in common materials can be better understood on a cellular level. The harvesting of lumber from trees also requires considerable effort to remove excess water, which is removed from the outside in an inward direction. Water exists within the structure of wood products in three forms: free water, water vapor, and bound water. As wood is dried, the bound water is the last to be affected by drying.[8]

Freshly harvested wood can easily have a moisture content of 30% or more, and building codes require lumber used for residential purposes to have moisture content less than 19%; the wood in furniture can be much drier yet. Moisture content is a percentage composition of wood that is responsible for properties of wood such as its shrinkage, swelling, and strength. This easily observable relationship of moisture in cellulose causes one to ponder the effects of fluid retention with synthetic polymers.

7.6.4 Dimensional Stability Issues

Issues that arise in the analysis and comparison of impression evidence have often been misunderstood or ignored. A small difference in the size of a crime scene marking, as compared to a suspect donor, would be considered a matter of interest in a scientific research project. On the other hand, attention to such detail would be problematic if time is of the essence in saving a life or there is some urgency associated with getting to the next scene. Dimensional stability of familiar substances is a fallible assumption. Changes, including swelling or shrinkage, for instance, are easy to understand when the effects are rapid. A gradual change in an outsole or tire is particularly difficult to imagine since it is usually a relatively small event in the life of products that one trusts and relies on daily.

Many of the products that we comfortably assume to be static are known to shrink or swell daily. Log homes are specially constructed to allow for the shrinkage of the logs over time. Many products are far costlier to ship wet than dry, and even friction skin can swell to the point that incipient ridges appear as wide as any other ridge (as seen with some drowning victims).

7.6.5 The Importance of Friction

It is accepted that the force needed to set something in motion is greater than that to maintain the kinetic state. Friction is a cause of wear between surfaces in motion that will oscillate without the intervention of some form of lubrication. A small amount of friction that would make little difference in a single contact can bring about significant effects with longer exposure.

The oscillations emanating from dry friction are well known. Dry friction appears as resistance against the beginning of motion starting from equilibrium (stick mode, a constraining force) and as a resistance against an existing motion (slip mode, an applied force). Research of friction-induced oscillation revealed both familiar cyclic behavior and the possibility of an accompanying chaotic motion.[9]

It is fairly easy to find examples of friction in our daily lives. Car tires make frictionally induced noise as they move. The on again, off again dry friction heard as train wheels navigate a curve is a vibrating sound that shares much in common with the resonations of a bow drawn across the strings of a musical instrument. Stick and slip are even more dramatic when observed as a geological shift in the plates of rock in the crust of the earth, which is referred to as an earthquake or a tremor.

The role of friction is further defined by its removal. Stick-slip includes, in slip mode (creep), a degree of friction loss, but where friction is not present in a kinetic system, its absence is impressed by the resultant lack of control over direction or deceleration. Examples of friction removal are typified by the sliding of a sharp skate blade on ice or hydroplaning while driving a vehicle on water.

The measure of friction is the coefficient of friction. The apparently simple relationship of force and motion on an object is affected by many conditions, such as temperature and the nature of the surfaces. It can be fairly difficult to determine precisely what is in contact at a given time. Remember that with footwear outsoles and tires, only a fraction of the rubber pattern is in a given area; further, the asperities on the surface of a substrate provide yet another limitation to any calculation of the area of real contact.

References

1. NDT Education Resource Center, Brian Larsen (Ed.). 2001–2010. The collaboration for NDT Education. Iowa State University. http://www.ndt-edu.org (accessed June 12, 2010).
2. Tomczak, M. 2002. Properties of seawater. http://www.es.flinders.edu .au/properties_of_seawater.html (accessed June 12, 2010).
3. Hamm, Ernest D. 1984. Cyanoacrylate, maybe. *Identification News*, 34(5), 5.
4. Johnson, K. L. 2004. *Contact mechanics*, Cambridge University Press, Cambridge, England, 242–243.
5. Reynolds, O. 1875. On rolling friction. *Phil. Trans. Royal Society*, London, England.
6. Rand, C., and A. Crosby. 2007. Schallamach wave periodicity in soft elastomer friction. *Bulletin of the American Physical Society*, 2007 APS March Meeting, 52(1). http://meetings.a ps.org/Me eting/MAR0 7/Event/58743 (accessed May 25, 2010).
7. Walczak, M. M., and M. D. Porter. 1992. Surface and interfacial chemistry. In McGraw-Hill, *Encyclopedia of Science and Technology*, S. P. Parker (Ed.), 7th ed., 18, 4.

8. Simpson, W., and A. TenWolde. 1999. *Physical properties and moisture relations of wood. Wood handbook: Wood as an engineering material.* General technical report. FPL. GTR-113: 3.1–3.24. USDA Forest Service, Forest Products Laboratory. Madison, WI. http://www.wo odweb.com/knowledge _base/physical_properties_and_moi sture_relations_of_wood.html (accessed June 20, 2010).
9. Popp, K., and P. Stelter. 1990. Stick-slip vibrations and chaos. Philosophical Transactions of the Royal Society of London A, 332, 89–105. http://rsta.royalsocietypublishing.org/subscriptions (accessed June 15, 2010).

Bias 8

8.1 Sources of Bias

Studies by Dr. Itiel Dror and David Charlton have proved a pivotal focal point for any discussion of bias pertaining to the identification process. While their research was conducted solely in regard to "fingerprint examiners," the implication is that any expert, in any discipline, is subject to bias. This fact should provide no solace for footwear, tire track, or tool mark experts.

The relative apparent simplicity of details within other forensic disciplines may be a negative influence in that bias is no less important to any of the disciplines. Few comparisons, in any field of study, are simple. When complexity in a comparison exists, so does the potential for error and bias. Allegations concerning the usefulness of such findings or the professional standards of participants must be shed in favor of understanding the vulnerabilities of all impression evidence comparisons.

The research included the actual casework of trained individuals who had received some level of acceptance in a sample of the identification and legal communities. The research was intended to be objective, and there is no reason to doubt that results of a similar nature could be obtained with other samplings from various locations around the world. This acceptance does not endorse the methods of testing as comprehensive proof but rather is a good starting point for the further research that should logically follow.

The potential for error is a common theme for any profession. "Can any other process boast such a reliable and highly accurate record of success? Certainly not the Health Service, where occasionally it has been reported that literally thousands of wrong diagnoses are sent out to unwitting patients."[1(p231)] Leadbetter's opinion regarding the vulnerability of decisions refers to the identification of individuals by their friction ridges. His opinion is illustrative although apparently somewhat defensive.

Attempts to rid a matter at hand of all but objective considerations are well intentioned. There is always, regardless of the efforts to control bias or error, the possibility that one or both can exist within a decision, conclusion, or process. The vulnerability extends from identification practices, scientific studies, and to research regarding the conducting of experiments designed to gauge expert decision making.

Let us consider the possibilities offered by the research of Dror and Charlton. The obvious conclusion, no matter what kind of misdirection is chosen, is that bias was discovered to have resulted from the introduction of context. The notion that the obvious answer of "any fool can see" is a common form of bias. Correctness may also consist of subtle layers or levels.

The conclusions of bias studies, to date, are entirely dependent on the use of context. Context studies into bias thus far lacked the use of an independent control, but the results supplied good reason to seek methods of reducing bias. Another vulnerability to error has not yet been discussed: Matter has to do with the path by which a complex pattern is analyzed, even when examined at another time by the same individual.

The path to decision making chosen by an examiner can be fraught with perceptional dangers. The way human perception is used may be considered in this way[2]: "We don't perceive the world merely from the sensory information available at any given time, but rather we use this information to test hypotheses of what lies before us. Perception becomes a matter of suggesting and testing hypotheses" (p. 222).

Personal experience with highly difficult friction ridge markings will verify the dangers posed by some impressions that test the limitations of our hypotheses to a high degree.

This line of thinking leads to the frame of reference provided by examples of sequential paths for comparison of friction ridges and other forms of impression evidence. The methods of approach are separated by the types of markings (biological and mechanical). Yet, the potential for error exists in all subjective analyses.

There are different scenarios in forensic work that illustrate common sources of occupational bias. Surprisingly, senior staff may be more at risk of certain types of bias than initiates. Imagine a supervisor who states, "My staff would not make such an oversight." This supervisor may have to defend his or her methods, and the training of the staff on a daily basis, and his or her trust could be well placed, but such an assumption can lead to dramatic problems and unwise training choices. This reaction may be a product of the adversarial environment forced on all law enforcement and related activities.

Some biases are triggered by past experiences or memories. Traumatic memories or associations are those most likely to surface and to influence observers of a crime scene. Some of those biases are easily rekindled.

The senses are an amazing gift. Think about the color red, and you conjure an associated image. The mention of blood causes your mind to recall not only color, but also that characteristic metallic odor. The feel of blood, especially when associated to the sticky, slippery experience resulting from a memory of a significant personal injury, is a pervasive thought.

Mention the word *spatter* and your mind turns immediately to scenes you have experienced that typify the use of a variety of markings, each indicating the direction and nature of events. Colors, especially red, have been used for centuries purposely to draw attention to a specific area of a painting or composition of a photograph, landscape, or interior design. It should be no great surprise that blood can bias our judgment of the relative value of evidence at a scene of crime.

8.2 Creation Theory Bias

Creation theories can also be a source of bias. Consider an example of a practitioner working a potential scene of a serial offense that includes a penchant (on the part of the culprit) for entering residences through obscure basement windows. This would be the first event to include crawling underneath a small porch. The window was witnessed as the point of entry to the residence. Lattice around the porch and a ground sheet under the porch were obviously disturbed.

The practitioner scrambles eagerly under the porch to examine the covered outer side of the window. It is reasonable to expect a point of entry to be a primary source of evidence, but it may be stress, a lack of attention, or repeated exposure to a particular kind of evidence that causes this trained person to focus so intensely on the point of entry. This is an example of searching for the evidence you expect to find rather than considering that the groundsheet should have been removed and examined prior to any treatment of the window.

8.3 The Awful Truth about Bias

You see in Figure 8.1(a) what appears to be a bicycle track. Few practitioners are familiar with bicycle tracks as they are not commonly part of a crime scene. This particular example was quickly made using a straight edge applied twice and bordered by manually applying the edge of the footwear outsole (Figure 8.1(b)) to either side of the parallel center marks.

It does not matter whether you accept my example of coincidental events as well constructed. The point is that at first glance you will see a bicycle tire track when one does not exist. It is not a challenge of your abilities; you would not likely be the first person on scene, and you know what a layperson would suggest as the source of this marking.

The challenge for the practitioner is to overcome suggestion and treat the marking with suspicion. On the other hand, you would not want to be in the position of having to explain the presence of that marking, seen within a photograph, in court without some attention given to attributing it to one source or another.

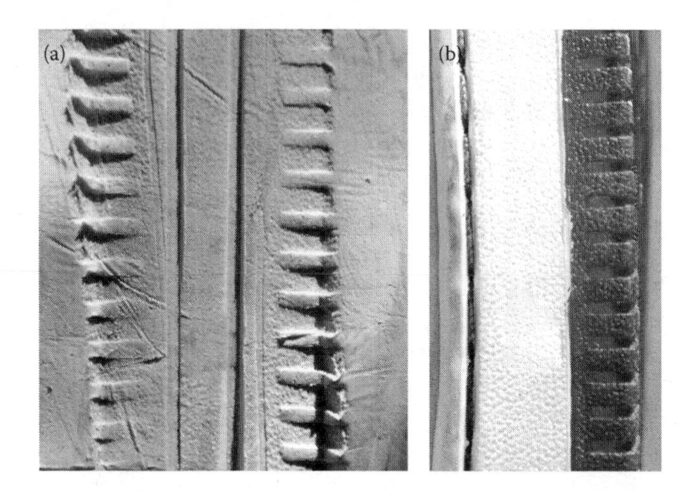

Figure 8.1 (a) This appears to be a bicycle track. (b) The edge of this footwear outsole was used to create the parallel notched elements on the left. You would not expect to find this at a scene but coincidences do happen.

Concepts of bias cannot be fully understood without some context. The truth hurts; it is often counterintuitive, and facing the possibility that one's decision might be biased is not easy. Studies of bias might easily begin with the mechanisms of prejudice—imagine an audience not introduced to the contents of this chapter.

Ask the question: "Are you prejudiced?" The responses will vary among yes, no, undecided, not sure, or maybe. Which group is telling the truth? The correct answer is "Yes."

The sources of prejudice are global. The form of prejudice is ever changing. Prejudice is omnipresent. Prejudice and bias are close relatives, introduced by our training, our contact with the world around us, and our daily experience of that world. Arrogance and ignorance exist within the extended family and are similarly intertwined.

If one accepts, no matter your skin color or other physical attributes, that you have prejudices, then the presence and mechanism of bias become much easier to understand. This is not an easy task since most people pride themselves with possessing varying degrees of liberal attitude. The fact is that our senses are focused on, and dedicated to, making discriminations among objects, surroundings, processes, and other people.

The ability to discriminate is a necessary tool for identification, and no matter which of the senses is used, the result must be a product of discrimination. The problem with discrimination, prejudice, and bias is not in their existence but rather how they are applied.

Think of searching for a person and beginning with the information that the individual you are seeking is known as "Uncle Bob." Truthfully, there are not many in our profession who would begin that search by including the female population of a database. While the decision to search for male suspects is understandable, it also incorporates the discrimination of gender, the bias of gender-based name familiarity, and the logical prejudice that of all the Bobs you may have encountered, none was a female who could be readily associated with "Uncle" or "Bob," let alone the combination. As a matter of consolation, a computer program that includes gender bias would fare no better.

For those who are now wondering what possible connection there could be to identification practices, consider that we develop less easily pictured biases on a regular basis. Habits are biases imposed by repetition. The choice of a path will often be made more on past success than applications of reason or logic.

8.4 Errors

Everyone makes errors. It is a daily occurrence. You may take a wrong turn, spill something, or use an unfortunate or inaccurate phrase, but errors regarding opinion evidence must be eliminated or at least controlled long before an erroneous conclusion becomes a problem. These acts are rarely intentional.

Errors can also occur through no fault of your own. It is possible for evidence to be sabotaged by events beyond our control, such as vulnerability in supporting evidence from another source. When you are the one giving evidence, even that thought is of no solace. The problem with making an error is that it is virtually indiscernible from bias or oversight and indistinguishable as an intended or accidental mistake, and all are equally damaging to credibility.

8.5 Validation

The act of seeing is an amazing process in which inverted images received by the eye are translated into acceptable representations of the world around us. The process is dependent on a bias, of sorts, in that the mind is continually responding to the stimuli of patterns in light, shadow, and color that are compared in a fragmented manner to the memory of an associated image or condition.

The mind regularly anticipates and connects visual fragments as a comfortably familiar or expected result. This is a natural phenomenon. The opportunities for bias in any visual system are common. The question is not "Is there bias?" The question is "How do we minimize bias?" with any system that relies on human perception. Perhaps from this perspective the concept of an error rate is not entirely without merit.

The "principles of individualism" state that all things are unique. The uniqueness must be accompanied by sequential agreement between characteristics of such number and significance that they preclude the possibility they could have occurred by mere chance. A further codicil bounds this principle, establishing that there are no unexplainable differences.[3] This model works for the most part but it depends on the ability of the practitioner to control or eliminate bias and accurately apply the language.

It has been suggested that dissimilarities cannot exist as either explainable or unexplainable. A case was lost due to a challenge of the language based on the concept that dissimilarity negates identification. There will always be lawyers challenging language rather than finding truth.

If a marking is dissimilar compared to another and that dissimilarity cannot be explained, then the ethics of the matter dictate the need to cut bait, so to speak. If the source is distortion and it can be explained as a recognizable form of distortion, then the two markings can be considered identifiable, and any dissimilarity between them can be explained.

Harold Tuthill, on Ontario Policy College forensic instructor, once asked his class to identify a footwear impression—they declined. There was dissimilarity across the width of the imprint that could not be resolved. This response was sound. Subsequent case work involved a similar issue: In a robbery, imprints on a bank floor were effectively divided across the width of the outsole by a band that was later attributed to the straps of coveralls worn by the suspect. The value of the class study became immediately apparent.

A case might be lost (hypothetically) on the basis of a challenge to the use of the words *clarity* and *distortion* in reference to a single marking. Introduction of some confusion over the simultaneous and contradictory descriptions would not constitute bias or a legitimate challenge of a conclusion. It would simply be another example of the subterfuge that can be injected into court cases.

The unfortunate aspect of the matter of confusion over dissimilarity is that the initial premise regarding the individualism of any impression (no matter the source) has been too often ignored. Whether one would care to trace the roots of modern practice to David Ashbaugh, Harold Tuthill, Roy Huber, Paul Kirk, Galton, or even Mark Twain, the result is the same. The voice of evidence is the source that needs unbiased and undivided attention.

It is difficult to imagine the vulnerabilities of perception without the existence of extenuating circumstances. Notable examples occur when the original characteristic is unique in form but is attributable to a damaged part of a mold or, with friction ridges, a temporary contamination of a portion of the ridges that causes the repetition of a nonexistent feature. Equally damaging to credibility is any lowering of your standard of acceptance that allows you to fit the facts to the sequence rather than finding sequential evidence that is acceptable.

Subsequent impressions that contain randomly formed mold characteristics can be virtually indiscernible from examples of typical postmold individual features. The only reliable method of detecting the distortion is to locate previous or subsequent mold impressions that either rule out or suggest the possibility of mechanical duplication in new footwear. A minor point of assistance is provided in that the appearance of protruding features in a newly molded product can be attributed to mold characteristics or damage, such as the sprue (filament-like artifacts that are easily damaged) seen commonly on new tires. These are formed by the gates of the mold through which the rubber product is injected or indentations on the mold itself from coincidental damage (see Figure 8.2).

These artifacts of the molding process may cause some interesting sprayed effects with liquids. And, because they wear quickly, may leave small stublike remains or be worn away completely, depending on the amount of use between making the markings and the examination of the tire.

Figure 8.2 Sprue, a significant but temporary feature of new tires that will generally wear away within days or even hours of use.

Mold deposits and foreign items dropped into the mold would form voids that can be much more difficult to detect. This condition can be largely negated with older footwear that shows sufficient wear to erase any potentially confusing molded imperfections. The proof can include observations of differential wear compared to new versions of the same footwear. Wear itself is a measurable feature for which the depth of tread on both footwear and tires can be used to indicate the relative age (in terms of wear rather than time). Shrinkage of components, cracking, and additional damage can also be used to substantiate the comparative age of footwear.

8.6 Confirmation

Errors can be produced by means of confirmation bias; this is the same area of concern that would include the power of suggestion. The significance of confirmation bias is the basic tendency of people to prefer to confirm rather than deny any proposition. In light of this basic tendency, it makes perfect sense that a preponderance of results would confirm a hypothesis.[4]

The introduction of an impression for confirmation cannot be entirely without bias. The mere act of requesting confirmation would need to be reduced to a normal office practice of including a series of impressions on a weekly basis, for instance, as a fun activity, regardless of the presence of a case requiring confirmation. This suggested practice would offer the added benefit of providing a skill-testing routine in addition to neutralizing the impact of performing a confirmation even if the casework is buried in a selection of control samples.

The actual sample group would become evidence. The sequence or method of sample presentation may become a matter of debate. Even the history of such a practice would offer some proof of an attempt to systematically eliminate bias.

References

1. Leadbetter, M. 2007. Letter to the editor. *Fingerprint Whorld*, 33(129), 231.
2. Gregory, R. L. 1973. *Eye and brain, the psychology of seeing*. 2nd ed. McGraw Hill (World University Library), New York.
3. Tuthill, H., and G. George. 2002. *Individualization: Principles and procedures in criminalistics*. 2nd ed. Lightning Powder Company, Inc. Jacksonville, FL.
4. Sternberg, R. J. 1995. *In search of the human mind*. Harcourt Brace College, Orlando, FL.

Exhibits to Evidence

9

9.1 Re-Creating the Scene

9.1.1 Considering Observations

The purpose of gathering evidence is to create an image of how related events transpired. Evidence that supports or refutes a particular hypothesis is no more or less important if it does or does not support a popular opinion. It is further possible that one hypothesis can contradict another while both have merit.

Consider a wealth of evidence that places a suspect gunman at location A. The evidence suggests a hypothesis that the fatal shot was fired by the suspect at that same location. Then, consider that the evidence regarding the fatal wound indicates that the shot could not have originated from location A.

It would be wise not to jump to a conclusion in a matter such as this. The next step in such a case could be to consider alternatives; perhaps a ricochet might account for the discrepancy. Another possibility is that the supporting evidence may rightly or wrongly cast doubt on a vital piece of information, such as a mistake in a pathologist's report that confuses an entry wound for another entry wound or even an exit wound.

Now, imagine reviewing a footwear case of a serial nature (more than one crime scene involving the same pattern) in which you are provided with a number of test impressions made with precisely the same type of outsole. You are asked to confirm that none of the test impressions corresponds to the crime scene markings. Your findings reflect a confirmation of that hypothesis.

That might be the end of the story, but a careful review of the case may produce a different result. As part of your confirmation process, you find that two or more of the test impressions were created by the same outsole (suggesting that at least one of the pool of shoes to be compared has been overlooked). In itself, creating multiple test impressions with a suspect exhibit is commonly accepted as a method of establishing that features are reliably present in subsequent imprints. This should not deter you from checking to make sure that all of the suspect outsoles are correctly represented within the collection of test impressions.

The examples given are actually run-of-the-mill tests of a practitioner's abilities. Complications introduced by impressions that may have suffered significant change in the interval between contact and comparison are much more difficult to handle. This type of complication demands more attention to detail, both with recognition of the condition and in subsequent comparison.

9.1.2 Preconceptions

The line between an expectation and a bias is thin. Many people are surprised by the size of their fingerprints as rolled impressions simply because their expectations stem from a simple direct (flat) imprint. Many more have been conditioned by popular lore and common experience to expect that some form of durable impression evidence exists at each crime scene.

The judicial system is no less immune to preconceptions. With the knowledge that every contact between one substance and another leaves a potential trace, it seems logical to doubt the proficiency of an examination or to level blame when no useful evidence is located. Complainants who feel violated or outraged will routinely have great expectations that do not include a realistic vision of impression evidence as the fragile markings that they are.

Experienced practitioners learn that it is easily possible to lose evidence by misjudging the durability of impression evidence. A few types of markings, such as etched friction ridge markings (most often found on glass) or tool impressions that cause a three-dimensional change in the substrate condition, are an exception. Even the polymerized markings that result from glue (cyanoacrylate) fuming techniques can be degraded by rough handling.

A practitioner hoping for fingerprint evidence from plastic bags will often select cyanoacrylate to perform noninvasive treatment of the surface. The problem is that some evidence could be lost just by making that choice. Markings, such as the patterning of surface chemistry on plastic films (see Figure 9.1), will be negatively affected by anything other than gentle treatment, such as a mist of blown powder that adheres to the chemicals. These patterns are obliterated by glue fuming or vacuum metal deposition.

The latent markings seen in Figure 9.1 were treated with black fingerprint powder. Surface chemistry markings require patient examination that may, in some cases, be used to provide evidence regarding

1. Similarities in the method of manufacturing (i.e., handling marks from rollers or grips) between two bags
2. A probability that a specimen originated from a sequence of bags (where markings are found to be cut in half)
3. Consistencies or similarities between the appearance of surface chemistry, construction, and size of two different products

It must be noted that these markings are extremely fragile. Exposure to liquids will disrupt or obliterate surface chemistry. The marks in Figure 9.1

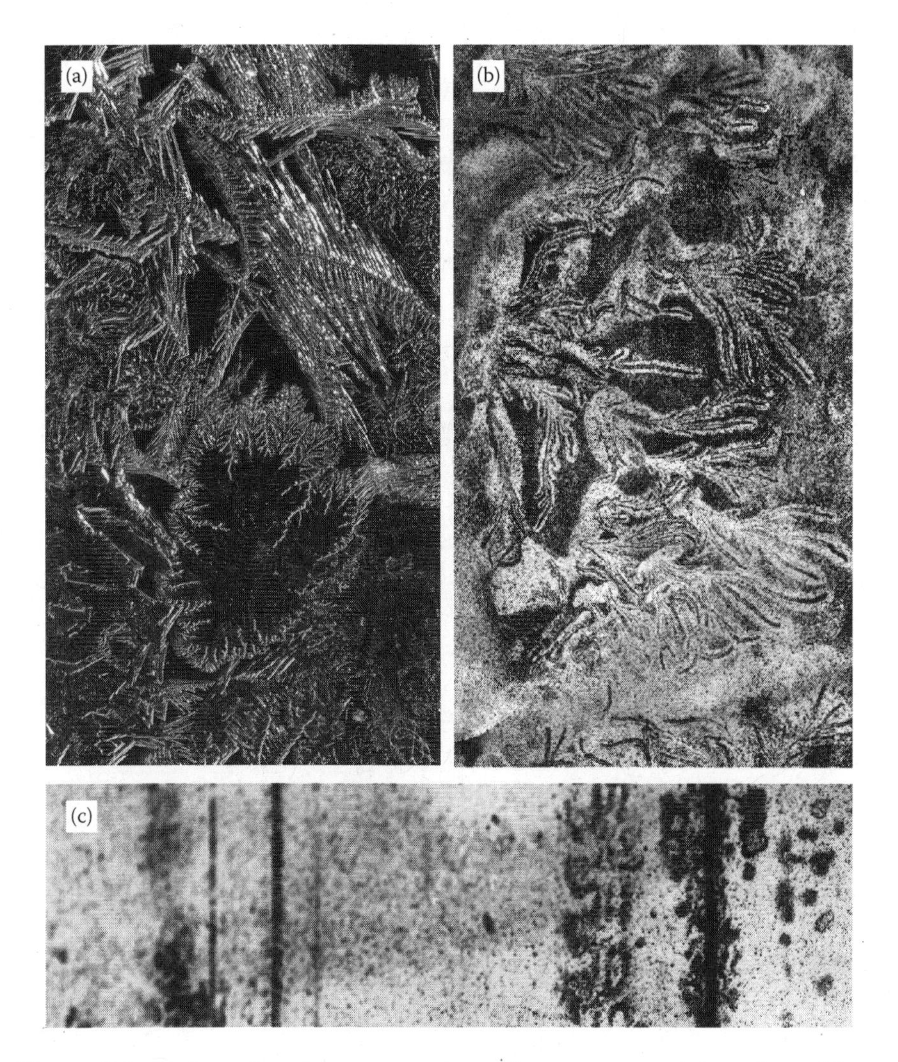

Figure 9.1 Surface chemistry patterning. Three patterns: (a) section of frost on a windowpane; (b) chemical patterning inside the gusset of a low-density polyethylene grocery bag; (c) roller markings in the same gusset as in (b). While markings in surface chemistry are formed by additives under the influence of pressure in the presence of air, the resultant markings can easily resemble some frost patterns.

were developed by a rather messy process in areas of the plastic that appear otherwise devoid of markings.

Chalk was too coarse to develop surface chemistry patterns. A coating of carbon smoke (subsequently exposed to compressed air) was an ineffectual development technique. Great care and personal protection is recommended to avoid both damage and injury, particularly when using powders with blasts of compressed air.

These latent chemistry markings are found on some plastic products and not others. The presence of additives varies in composition from one product to another, as does the appearance of the corresponding markings. The usefulness of looking for this type of evidence is restricted to only an occasional opportunity when new or nearly new plastics are examined. Despite limitations, this is yet another example of evidence that is seldom considered and often missed.

Preconceptions are best avoided by treating each specimen (piece or exhibit) of evidence, no matter how similar in appearance, as a separate entity. This is best done by making a choice of treatment based on a list of observations that is as complete and thorough as possible. These initial observations (e.g., the marks appeared as small, roughly circular outlines; the presence of surface chemistry was determined to be unimportant to this case) contain only statements of fact with no embellishment; they form the beginning of an analysis and allow anyone reviewing your actions to see what you did, what you saw, and why you took the course of action you chose.

9.1.3 Condition Theories

Condition theories (dealing with the specific conditions that support or refute a creation theory) can be a vital investigative tool but are not complete without consideration of every available observation. Consider a creation theory that suggests the presence of moisture at the time a contact occurred, which in itself seems mundane, unless the substrate is located in a desert or protected from the elements. Condition of a substrate and how that condition fits into the equation of an analysis of the exhibits is important for establishing subsequent estimates regarding the value of the evidence (and the accuracy of a creation theory).

An account of a condition theory must be based solely on reliable observations. Observations may include the following:

1. The size and shape of a piece of glass are noted.
2. The surface has roughly circular markings in the form of outlines in sizes between 1 and 5 mm.
3. Markings are evenly dispersed but only on one side of the glass.
4. The markings appear to have been developed by an application of black powder (if examining for verification purposes).

All of these observations may later be used as the basis of a condition theory, such as “The small, roughly circular markings developed with an application of powder on the substrate appeared consistent with the size and shape of raindrops at a given angle." This may be further supported by other observations or even independent information such as a weather report for the given area that may or may not describe potentially consistent events or situations.

It is common for a theory to be based on a number of observations. Preconceptions, such as merely jumping to the conclusion that the markings are raindrops without the basis of prior observations, can lead to bias. Remember that, in fact, the observations led to a condition theory that needs the support of more data (how can you be sure that the marks were made by water?).

Each tidbit of information contained in a list of observations can be accompanied by remarks that constitute condition theories, which in turn are followed by creation theories, all of which serve to build a foundation on which to finally state a conclusion or opinion. These painstakingly thorough methods can pay dividends in and out of court.

9.1.4 Creation Theories

Whether cognizant of the act or not, an examiner will naturally develop one or more creation theories as an examination of a scene or an exhibit is performed. Creation theories are a set of observations that indicate how an event may have transpired to cause the observed effects. It is typical of a first responder to classify an event in such a way that a potential for bias is created by the first-available creation theory, which can now be defined (whether it is ultimately right or wrong is irrelevant) as a preconception.

Judiciary systems are easily as vulnerable to developing preconceived creation theories based on the evidence heard. These deficiencies are hardly intentional, but both situations can have an equal, if opposite, negative effect. It is never inappropriate (for instance) to consider that a judge or jury knows nothing whatsoever about the fragile nature of impression evidence and in fact may have been misled by media accounts about the reliability and durability of each new forensic discovery.

A broken window with glass on both the inside and outside and signs of activity on both sides of the opening does not exclude the possibility that the artifacts may only be the result of a single passage. Even a preponderance of glass on the outside may simply be attributable to a small opening inward from which a tool is used to break the remaining glass outward. Supportive evidence must be located that confirms a particular theory.

A successful confirmation could include a suspect's footwear imprint under broken glass on the inside. This theory also reinforces the need for good scene management to ensure that the glass over the footwear could have been distributed after the fact, thus negating the likelihood that the window had only been used as a point of exit. A thorough examination of a scene where there is any doubt regarding the entry and exit may include noting examination of ceilings, crawl spaces, other attempted points of entry or exit, and the efficiency of locking mechanisms on all openings or accesses.

Impressions, even on a single exhibit, will provide multiple creation theories. Records that include those theories and exhaust the unlikely events by means of logic are valuable in court. Overstating or attaching undue value to a theory is to be avoided, but personal research or seeking the advice or assistance of more qualified sources should be encouraged as it would be in keeping with the verification of evidence (ACE-V, analysis, comparison, evaluation, and verification).

Creation theories are applied by engineers in studying areas of contact, such as the line of contact seen with a roller bearing as opposed to a single point of contact (as found with a spherical object)[1]:

> A theory of contact is required to predict the shape of this area of contact and how it grows in size with increasing load; the magnitude and distribution of surface tractions, normal and possibly tangential, transmitted across the surface. Finally it should enable the components of deformation and stress in both bodies to be calculated in the vicinity of the contact region. (p. 84)

The implications of creation theories are, first, that one can theorize and test those theories against the evidence available; second, that where a contact occurs that leaves a distinguishable marking, a theory of conditions at the time of contact can be calculated. Calculations would need to be expressed in terms of thresholds that define upper and lower limits for specific conditions that can be supported by the findings of other relevant disciplines, such as mechanical engineering. This represents a different approach in the consideration of markings as evidence and the evidence that can be derived from markings.

Engineers would, at this point, insert a flurry of formulas and mathematical assertions to prove the point and illustrate the boundaries of some practical thresholds (unfortunately, an in-depth analysis lies, for the most part, outside the scope of this book). Visualizations in mathematics offer a valuable tool in the assessment of a particular event in terms of the topography, elasticity, and change as imposed by variations in load. We can imagine a tire under load, which with change in air pressure as the load is applied, will bulge the sidewalls and press deeply into mud causing an upthrust of soils affected by the pressure. This can be modeled with mathematics.

Creation theories can be expected to vary significantly from one substrate to another and from one discipline to another. Friction skin will not demonstrate the same flexibility as footwear outsoles or tire treads, yet they are not that far removed from each other in terms of basic elements. Certainly, each discipline includes matters governed by similar elastic responses, which may benefit from applications of mathematical modeling.

The primary difference between forensic practices, then, is the formation of theories, opinions, or conclusions. A representation of a likely point of entry to a scene or the handling of an exhibit is an opinion that is viewed in court as an expert's conclusion that, with sufficient exhaustion of alternate creation theories, may be tendered as opinion evidence. "The precision of a numerical result is concerned with its reproducibility when measured again with the same instrument and is therefore an expression of the uncertainty due to random error; the accuracy of the result is an expression of the uncertainty including that due to systematic error."[2(p18)]

The exhaustion of possible alternatives is a common theme in forensic evidence and could be treated as though it were a numerical entity. The more problematic attributions of identity or isolating the source of a marking are not so easily managed.

Problems with the details of a tool mark, the lands and grooves on bullets, or other such striated imprints arise from the variety of pressures, forces, and angles involved in the creation of the markings. While the bullet striae are now routinely searched against a database and algorithms exist that allow one to "straighten out" wavy or irregular marks,[3] these measures only improve the accuracy of subsequent analysis and comparison. Straightening a mark is only a part of the solution for the larger sources of striations.

An ideal solution would include software that could interpret any type of marking. This has been approached by more than one individual with varying degrees of success, but no commercially available version for tool marks is available yet. The advantages of such programming would be the removal of human error from related equations and the automation of searching both fine and complicated patterns.

Exploration of current forensic workplaces show how few of the individual disciplines include research as part of the work environment. This is a management issue that has a great impact on whether an employee or the courts define tasks as science or merely a practice. This lack of definition is both lamentable and exciting since it betrays how much remains to be learned and explored and, further, why it has not developed to full potential.

9.1.5 A Closer Look at the Role of Observations and Theories in an Analysis

The first role of notes is to record the observations. If the person doing an analysis also attended the scene, those initial notes provide a foundation for the more intensive observations of an analysis. Laboratories that conduct an analysis of exhibits removed from the scene may also require a copy of notes and other records from the investigator for referral in accordance with their protocols.

The analysis usually takes the form of a report containing a brief case description and including details that pertain to the evidence in question. Observations will outline not only the state of an exhibit and any visible features but also the presence of anomalies, matters that pertain to the location in which it was found, description of any treatments or exposures that might have had some effect, and finally its measurements.

The initial use of listed observations is not to form conclusions but rather to establish theories regarding conditions. Returning to an examination of a single piece of glass, the report (verification phase) may begin as follows:

> On a given date, pertaining to a particular case, an examiner was presented with a single piece of evidence to be verified. Exhibit 1, details of which were recorded in a set of bench notes and observations, were provided.
>
> **OBSERVATIONS**
>
> Exhibit 1 is a shard of glass of irregular trapezoidal shape measuring 11 mm along the smallest edge and, in clockwise order, 36, 28, and 42 mm along the other three sides. The glass had apparently been dusted with black fingerprint powder. Markings were observed, interspersed on the surface of the shard; these included roughly circular shapes that did not appear solid but rather took the form of outlined features averaging about 1 mm in diameter. Fine-patterned, wavelike markings that appeared to be arranged parallel to each other, with an average width of 1 mm and 1 mm apart, were observed at the narrow end of the shard (marked with a designation of F1 in China marker), and what appeared to be partial friction ridge markings (marked with a designation R1 in China marker) were observed about halfway along the shortest side.

The language and descriptions of the previous paragraph are likely to be considered thorough and would certainly be enough for most practitioners to roll their eyes. A description that needs such depth is usually nicely covered by one or two images. The point in looking at the wording is that in court you may have to describe your observations, which means that you need to hone an ability to put observations into words.

There are a couple of problems with all of this preparedness. The first problem is that, despite wanting scientific evidence, most courts are interested in not much more than the bottom line. A practitioner must somehow find equilibrium between what can be done and what should be done.

The other problem with the example observations is that they lack attention to many basic details, such as the thickness or color of the glass. The description further mentions only one surface of the exhibit; it was broken with no description of the broken edges or if one or more edges had not been broken. The pattern was not measured in terms of the amplitude of the waveforms, frequency of waves, presence of voids or anomalies, or even the number of waveforms counted. The area described as "possible friction ridges" might have been referring to the muzzle print of a dog since the reader is left to ponder why it is only suspected of being derived from friction ridges. The description of what could be some dried liquid is not referenced with respect to either of ridge detail or the patterning.

This attention to detail must extend to any suspect exhibits, such as footwear, tools, tires, test impressions, and debris. Reports need to employ concise language that conforms, if possible, to both common understanding and subsequent observations. Take, for example, a description of a piece of fabric designated as having a herringbone pattern. The term *herringbone* conveys an image of the general pattern, but when adapted to fit a description of a footwear outsole pattern it does not fit well or assist with subsequent details.

The use of more concise terms and unified descriptions is not a substitute for evidence, but it will improve the results of casework reports when applied. Describing the evidence at hand in a complete, rather than abbreviated, version may make the difference in establishing the credibility of your observations.

9.1.6 Comparison

Armed with a battery of observations, condition theories, and creation theories for each specimen (exhibit) and each suspect tool (whether a shoe, hand tool, or inked impression), the practitioner is now ready to sift through the data, shaking the evidence tree to see what falls out. Remember that up to this point all of the theories and observations were not comparing one thing to another but merely analyzing each specimen to determine a baseline of appearance, condition, and description.

A good place to begin a comparison is to organize the various elements to be compared. As suggested in this chapter, it is wise to ensure that test impressions correspond to the suspect exhibit. Personal taste will dictate exact methods; some prefer the use of preprinted report forms or evidence lists, but one should remember that whatever the choice, it must be designed to comfortably translate into comprehensible testimony as required.

Think of impressions as originating from a mixture of conditions. Mixtures, by definition, contain more than one ingredient, and the creation of an impression must be considered as a mixture of ingredient circumstances. Changing the way in which evidence is analyzed and compared to a consideration of ingredients and results of the combination of those ingredient circumstances would be a radical shift in the way evidence has been traditionally considered.

The benefits of treating impressions in such an elemental way stem from a neutralizing approach in which each hypothesis is considered, including, or perhaps beginning with, the antithesis. A statement of opinion becomes much more compelling if treated elementally and is further enhanced by legitimate consideration of the opposite hypothesis. In other words, in the matter of comparing two markings, try proving that they do not correspond, and that the suspect (or donor object) could not have caused the impression.

This is a method of applying the principle of "establishing that there are no dissimilarities." The task begins with the comparison of those observations of details, namely, the measurements and descriptions (particularly, creation theories in which the specimen in question is likely showing similar or dissimilar signs of wear). The next step is to locate those features or areas that do not coincide or coincide poorly.

If no adequate creation theory applies to the areas of dissimilarity, one must head back to the drawing board, so to speak, and take a closer look at the evidence and any test impressions that might contain further information. Remember that a negative hypothesis can still be essential evidence and is worthy of the same scrutiny as any other piece of the puzzle. This is especially true when testimony of a negative hypothesis, stating that an impression was not made by a particular source, may contradict popular opinion or open a new line of investigation.

Imagine two waveform patterns in a marking, of the same general description over a single wavelength, that provide a total count of six waves; the last two merge into an area that resembles the worn heel of a shoe. A test impression records similar detail in similar areas, but the waveforms are somewhat wider, and there is evidence of another parallel but intervening wave. Pressure can account for just such an event but so can a different outsole; you now have a considerable amount of work ahead of you to determine which scenario is correct.

9.1.7 Cost Factor

Reality for most practitioners is a matter of cost, either in time or in money. Paying the amount of time and attention required to nail down details as described here requires an expenditure of both for supervisory staff who want their lab or unit to reach a higher level of achievement. Practical considerations will most likely play a role for most practitioners in deciding which cases receive full treatment.

Investing time and effort in reports can have the effect that fewer cases will go to trial; the reason for that is twofold:

1. Lawyers, faced with an improved explanation of the evidence, may actually be able to assess the reasons that a guilty plea is the most reasonable option.
2. The weight of the evidence, provided that it remains clear, will only improve an understanding of its relevance to both judges and juries.

9.1.8 Conclusions

Improved measurement of features within impressions amounts to only one factor in building a thorough treatment of a case. Case management will have a direct impact on the quality of evidence. Consider a report that lists observations regarding a case for which subsequent tasks have been completed by separate individuals.

The reason for the specialized division of duties can derive from the best of intentions and still disrupt the flow of converting observations to evidence. Scenarios that point out the greater number of potential sources of exhibit contamination or the added difficulty in maintaining a chain of evidence are fairly easy to imagine. Other details are less obvious, here are a few theoretical problems to consider:

In a theoretical case, first responders have created a taped boundary around the scene of a homicide. The scene boundary includes the victim's car in the driveway, the front porch with a broken front entry door, and a blood trail leading from the back of the victim's car into the house. On arrival, a team of two forensic practitioners photograph the scene, then examine the crime scene log to see who attended inside the barrier and at what time. They are then briefed by the first responders that a vehicle (according to a witness's description) had been seen pulling into the driveway behind the victim's car. The lone occupant of the newly arrived vehicle and the victim entered into an altercation, from which the victim fled into his house. He was followed by the suspect, who kicked the door open; shortly after entering, the suspect ran from the house and sped away in the same vehicle that he had driven to the scene.

A fingerprint is lifted from the driver's door of the victim's car, a tire imprint is photographed in the paved driveway, a pry mark is photographed on the entry door, and a partial red-stained footwear imprint is photographed on the exterior of that same door. Assume that each of the impressions was suitably photographed, each provided prolific details, and that there were no problems with later obtaining the corresponding exemplars from the suspect, his fingers, his clothes, his shoes, his car, and his switchblade. Think about the vulnerabilities of the case if each piece of evidence is subsequently forwarded to a lab where the examination is continued.

Assume that there are no problems with procedures or the chain of evidence, the known samples were all legally obtained, the laboratory is well accredited, and the treatment of the evidence is appropriate. In fact, the only weak links exist in how the scene was handled and the level of communication between the parties doing the investigation. Now, you might be able to pick out at least one or two possible flaws.

Tire impressions photographed on the paved driveway: Tire evidence that does not include some element of depth is a two-dimensional marking. Markings on pavement can be identifiable, but they are in competition with every other tread that uses that same driveway. There are a few areas of concern that are often overlooked with such evidence:

1. A long driveway in particular is seldom protected from incidental contact by emergency service vehicles. Test imprints and an account of the vehicles that were known to attend the scene should accompany any lab submissions.

2. The victim's vehicle is often overlooked when test prints are made, often due to the knowledge that it was followed into the driveway.
3. Unlike other forms of evidence, tire imprints are often disregarded by well-meaning investigators who do not afford such evidence the same priority treatment as more obvious or more impressive markings, such as a blood trail.
4. Few investigators are prepared, in any way, to preserve openly exposed tire evidence from weather or any other form of degradation.
5. The general public, distraught family members, prosecutors, firefighters, paramedics, and even some investigators are proud to claim that they have learned so much from watching television. They are excited to see what happens next, even when the next step is to prevent them from stepping on any more of the evidence and obey the message printed on the cautionary boundary tape.

Footwear impressions photographed on the outside of the door used as the point of entry: Particularly where the imprint appears to have been made in blood, these are more likely to survive the onslaught of spectators when scenes are not well preserved. These marks may appear in fairly pristine condition but are more often not without some damage. Vulnerabilities include

1. Contributory patterning, for which the matrix came from a source that may well have been altered by the victim's footwear patterns prior to the suspect contact, may result in the transference of a pattern within a pattern.
2. Distortion often accompanies a marking made at an angle that does not conform to a normal angle of incidence.
3. Measurements of height are often taken from the top of a door jamb and may not account for a difference between the jamb and floor level, which represents the location of the planted or supportive foot during the execution of a kick.

While fascinated by the located marking on the door, some practitioners will forget to properly examine the planted or supporting marking (usually located on a floor surface), which may assist with analyzing both the marking on the door and the type of kick used, whether it seems to have been produced by a front kick (thus showing the greatest amount of detail between the ball of the foot and the toe), a side kick (characterized by a generally horizontal, flat, edge, or heel-dominated appearance), a heel kick (which shows some dominant heel characteristics), or an angular blow (which usually results in the generation of some striated features).

9.2 The Use of Observations

9.2.1 Transforming Observations into Opinion Evidence

The concept of incrimination has (or should have) little to do with the initial examination of a scene or an exhibit. Recorded observations are vital to proving the absence of bias and fortifying the value of subsequent comparisons. The results may suggest incrimination, exoneration, or an indication of an entirely different nature (accidents or simply proof of coincidental and unrelated events).

To form theories, one must rely on the accuracy of the records of the scene and exhibits, including any measurements, calibrations, or comparisons. The value of any conclusion must show repeatability in the form of verification by a colleague who is sufficiently trained to proffer it. This is the essence of the ACE-V methodology, which has gained international acceptance by forensic examiners.

Forensic observations, in essence no different from any other observations, however, are bound by many parameters beyond the nature of the subject. They must contain only what is acceptable as evidence in a court of law; they must be free of complicated language or jargon (which would only confuse a judge or jury); they must not include introductions of a past record of the accused; and they must only include those matters at hand or be admitted on the basis of case law. The value of forensic observations is in many respects tied to the accuracy of the translation of scientific observation into common language.

Scene management can easily be the first thing to go right or wrong about forensic observations. It only takes one individual to drive in, step in, or touch the wrong place to wipe out good evidence; this happens with alarming frequency. That the individual source, with the best of intentions, may be an emergency care provider, firefighter, police officer, or complainant is irrelevant to the result, and lack of scene management is at fault.

Scene management is often not the responsibility of forensic examiners, but a management issue can be particularly detrimental to the process of accurate note taking. Record keeping is one of the primary concerns of forensic examiners. It must be realized that the necessary records may need to encompass confirmation of events that transpired before the examiner reaches the scene or receives an exhibit. Events that are mismanaged may also consist of events that are not recorded.

This part of the task is akin to the observational phase of any scientific experiment. The difficulty posed by a crime scene or any impression is that even with the most routine cases, there is always the possibility that things are not as they appear, and the examiner must be prepared to respond and analyze appropriately. Crime scenes seldom allow little in the way of independent variables (influences or conditions that can be controlled in an experiment or at scenes and with exhibits by the examiner).

The use of independent variables is even more important when the choices are limited. Photography and lighting are two such variables that have historically been used to good effect; other variables such as temperature, time, and accessibility are a few examples of variables that are generally much more difficult or impossible to alter. Simply obtaining a record of the scene, however, is not enough.

Wear is an observed characteristic that many practitioners rightfully treat with utmost caution. Wear patterns can be described as either class (belonging to a subset of features that share a common origin) or (occasionally) individual (random or unique features). Wear imparted by a particular inclination of the gait pattern may be inconclusive when compared to the wear patterns exhibited by another individual.

The use of the word *inconclusive* seems to infer that there is no basis for further comparison when in fact it means that the comparison is inconclusive only when compared to another specimen of the same size, shape, and pattern that exhibits approximately similar amounts of wear in the same areas. Consider one specimen that matches the brand, model, mold, size, and lack of wear (as seen in a relatively new shoe) observed with the crime scene impression. The test impression taken from that specimen will undoubtedly be inconclusive when compared to the marking from the scene and, further, decisively conclusive when compared to other models made by a different mold or another brand.

There are reasons that wear is given a wide berth in forensic evidence. One is that the frequently asked question, "Isn't it true that there could be hundreds or thousands of [shoes, tools, or tires] that exhibit this amount of wear?" has no basis for any reply other than, "Yes, this example of wear is considered a class characteristic." Our comprehension of the formation and occurrence of the mechanics of wear lacks a scientific basis in fact. Understand also that there are no known reliable studies of sufficient magnitude to support or refute the concept that wear is not unique.

Unique wear can be attributed to the highly random generation of forces at play, such as exposure to a particular substrate or wear that has resulted, for instance, in some break in a pattern. The instance of unique wear in those rare instances when it does appear must be a measurable feature to be deemed random. In terms of the future, studies are ongoing that may one day offer more useful information about this topic.

For the benefit of anyone who either is required to give evidence about features of wear or is considering the furtherance of research about the subject, it should be noted that wear as a condition is measurable. The amount and quality of measurements may well provide the means, with a sufficiently robust study, of establishing the unique lineage of wear. Each pattern is generated by the different paths of each shoe as worn by different individuals. Where the measurements show sufficient correspondence, it may become possible to support the hypothesis that another shoe could not, with a high degree of probability, have made the impression in question.

Other possible results of simple experimentation are as follows:

1. Definition of the tolerance of a single given specimen to retain a precise amount of wear over time (usage) before a change occurs
2. Effects of introducing another specimen of the same source to precisely the same application of force
3. Effects on the stress-strain relationship as a result of differing bulk characteristics between two specimens

The conditions that produce the transfer of patterns or examples of wear require as much caution as interpretation of the markings themselves. We tend to imagine a suspect shoe, tire, or finger as a stable object even though we seldom have background information that would suggest reliability. Chemical exposures or applications of forces have been removed, yet the finger may be swollen or the rubber in an outsole or tire weakened.

The sensitivity of many situations may also require some attention. Forensic casework can be embroiled in local politics or even clashes caused by the perceived role of responders. Extraneous considerations such as the susceptibility of evidence to weather or court-related time constraints are not typical of research conditions as would be expected to affect other sciences. These and other constraints define a real separation between forensics and other professions.

9.2.2 Merging Technology and Practice

Interpretation of three-dimensional evidence (casts excepted) is often presented in a less-than-efficient manner. Consider that many three-dimensional impressions are converted to a two-dimensional form for comparison and charting. The change from three- to two-dimensional models, although convenient, is destined to result in some loss of detail, such as the profile of an extruded feature (see Figure 9.2). The conversion further challenges the ability of judges or juries to visualize the full impact that the evidence would have had prior to flattening.

Extruded materials are shaped by the edge of a harder material pressing against the yielding bulk of a softer material. The extrusions are subject to change as the harder material changes. Extrusion is an example of a process in which a softer material that contains abrasives will eventually be responsible for changing the harder material that marks it.

It is commonly accepted in scientific circles that a conversion from one form of measurement to another, even when carefully performed, represents a potential source of error. Evidence that must be translated between two- and three-dimensional forms can result in some unwanted confusion.

Figure 9.2 Extrusions. Multiple examples of modeling clay that has been extruded by the same outsole. The details on the outsole of the shoe could be expected to change in response to a great amount of extrusion (over a matter of hours or days perhaps) and remain unchanged by the small exposure when creating multiple test impressions.

9.2.3 Limitations of Current Practice

Current practices offer little guidance for explaining the appearance of a marking or the forces that cause markings of the sort to be generated. The limits of current practice generally restrict a witness to some poorly founded attribution to distortion that seldom contributes useful information about the conditions that need to be present to achieve a particular state or to exhibit certain features.

Applications of structured light (laser scanning technology, as discussed in Chapter 5) afford a highly efficient method of imaging impression evidence while providing the means to effectively measure artifacts and details on each available surface. This is a technology that will one day be merged with our development in understanding the mechanics of contact in that both surface contours and profiles of the topography in any chosen line can be illustrated.

The procedure of casting impressions has, time and again, proven effective at capturing detail, yet there are drawbacks to the use of casts. They are time consuming to create, particularly if many impressions are concerned or the temperature drops. They are difficult to store, and transportation and management of a number of casts can be unwieldy in court. There are alternatives suggested in Chapter 5.

9.2.4 Expanding the Scope

A common saying is that it is human to err. It is no less human to overlook, for whatever reason. Discoveries and inventions are born of issues that have simply been previously overlooked.

A substrate is a material that can support some amount of transferred detail that will relate directly to the nature of the contact. In examining a contact, the exchange may consist of a fairly predictable result between two differing materials, mechanisms, substances, or organisms. Exceptions, however, are often found in crime scenes where the mechanics of contact do not fit expectations.

In terms of design, one must begin by generating a vision of what properties an object or system should have. The next step involves devising a large number of possible solutions for attaining the desired characteristics. The final solution benefits from comparison to rival concepts.

Take, for example, an ideal replacement for casting techniques. We know that a point cloud could offer definite improvements to presentation, measurement, and storage problems, but other features could also be desirable. A computer model of three-dimensional impressions, used to re-create the formative conditions and perhaps offer determination of the direction and nature of forces at play, would best be illustrated as an animation.

9.2.5 Reduction Techniques

There is an axiom that is well worn by artists and designers: "Less is more." In fact, there is much truth to be had by achieving more with less. The design or process used will usually benefit by a minimalistic approach.

This can be true of impression evidence as well. There are situations that benefit from reduction. The origin of this particular idea stems from negative reduction, by which silver nitrate could be reduced on the surface of film that had been overexposed to reduce the overall density of that specimen to a more useful level. The same type of concept can be applied to a marking in grease or ink, for example (see Figure 9.3).

The substrate chosen for Figure 9.3 was glass microscope slides coated prior to use with either ink or axle grease. Preparations of this sort were used for all four examples. The first specimen (Figure 9.3a) was a control sample impression in grease. The second example (Figure 9.3b) was a friction ridge marking in ink that was covered with a second layer of ink, then subjected on the right to a spray of glass cleaner while covering the portion on the left with paper during spraying.

The third sample was a similar marking made to overlap the juxtaposition of two edges of glass slides covered in ink. The lighter portion (on the right of Figure 9.3b) was sprayed with glass cleaner at another location. The last example (Figure 9.3d) was treated with a secondary layer of grease, as

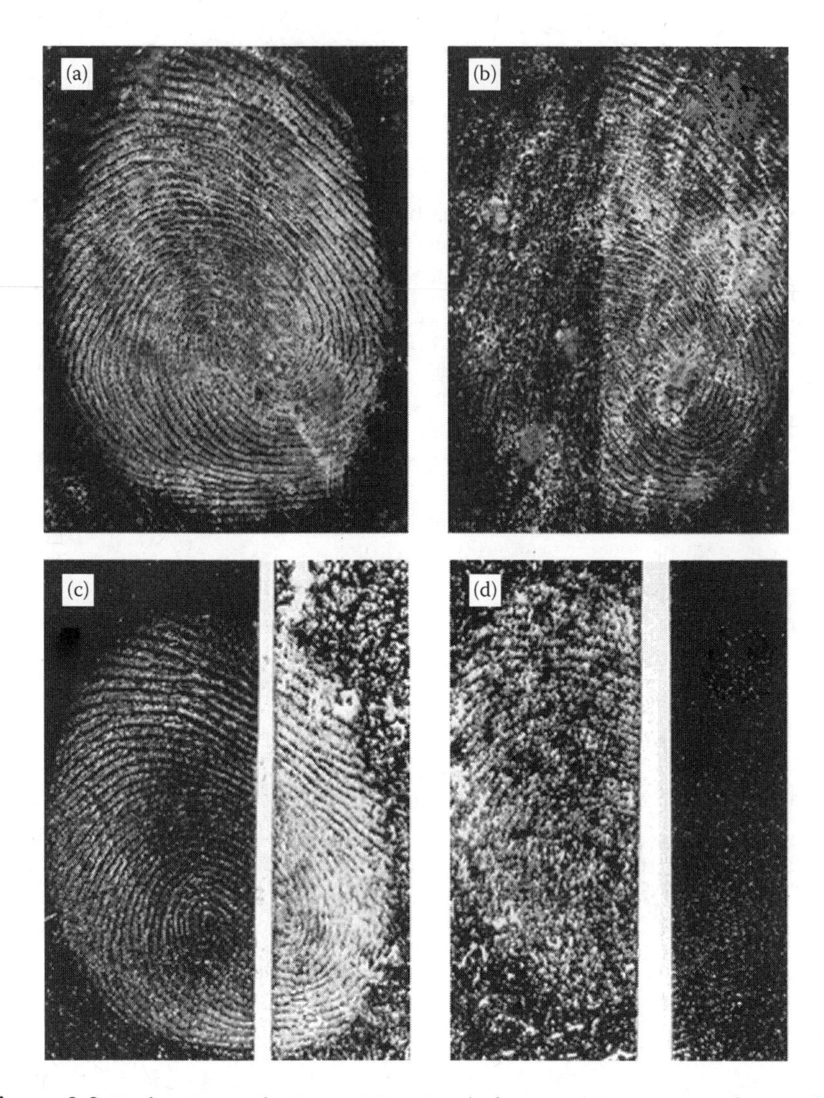

Figure 9.3 Reduction techniques. Negative (take-away) impressions from inked or greased surfaces. Impressions (a) and (d) were made in a thin layer of dark green axle grease, while (b) and (c) were placed in ink. (a) A control take-away marking. Glass cleaner was used with (b)–(d).

seen in the untreated portion (right portion of Figure 9.3d), and the other portion (left side of Figure 9.3d) was sprayed separately from the portion on the right.

The technique is not terribly reliable; it is subject to quick degradation from overspray and would need refining to be of practical use. The observations are interesting as a potential tip for dealing with markings that may be enhanced in this manner. Footwear impressions, tire tracks, or even tool marks may also benefit from this treatment but only if there is a surplus of markings for experimentation that would not be required as evidence.

References

1. Johnson, K. L. 2003. *Contact mechanics.* Cambridge University Press, New York.
2. Shoemaker, D. P., and C. W. Garland. 1967. *Experiments in physical chemistry.* 2nd ed. McGraw-Hill, New York.
3. Heizman, M., and L. F. Puente. 2001. Automated analysis and comparison of striated toolmarks. Presented at the European meeting for shoeprint/toolmark examiners, Berlin, Germany. http://www.ies.uni-karlsruhe.de/download/publ/hzm/sptm2001_pp.pdf (accessed June 27, 2010).

10 Validation Study of Three-Dimensional Striations from Outsoles

10.1 Beginning with Markings

10.1.1 Striation Marks

The word *impression* implies a vision of a relatively static transfer of details from one surface to another. The introduction of a three-dimensional aspect further conjures images of treads in mud as transferred by footwear or tires simply because the human brain works in an associative way, linking the sequence of words to what is most familiar. Striation marks are products of a dynamic situation created directly as the extruded product of compressive forces (with common three-dimensional impressions) between two or more surfaces.

These markings can be adjacent to the area of soft substrate contacted by the sole of a shoe and appear to have more in common with a bar code than a footwear impression. Striated markings (of any description) resulting from dynamic linear contact will retain some features that can be associated with a pattern, but with a difference. The markings caused by such a pairing will most generally appear as parallel markings, and the edge or surface that creates the marks can be regarded as a tool.

The effect commonly occurs where one surface is either impacted or abraded by another material, which may be of a different hardness or coated with a substance that offers a record of the contact. The characteristic relationship of a softer substrate to a somewhat harder outsole is accepted here as a starting point for validation studies.

10.1.2 Experiment to Define Striation Repeatability

The experiment discussed next was designed to study the presence of striations from a new state. The experiment was extended to examine the development of wear, by the introduction of controlled abrasion, on the same, otherwise new and unused, shoes.

10.1.3 Equipment

- Three new shoes of the same size and pattern
- A clamp and frame
- A wheeled platform

- Modeling clay
- A clay-forming device
- A permanent marker
- Digital calipers accurate to ±0.01 mm
- A digital camera
- 150-grit emery cloth

10.1.4 Methodology

The three shoes were purchased from two sources in two different models. Each of the shoes was clamped at a fixed position as marked on the outsole. Modeling clay was formed into 10-mm bars or cakes, then mounted on a wheeled device. Each shoe was then attached to a frame and impressed into the clay. The clay was forced across a selected portion of the edge of the outsole. This caused features from the edge of the outsole to be extruded into the softer substrate. These control samples were stored and photographed.

The next action involved the emery cloth, which was clamped in a strip about 180 mm wide and 200 mm long. The shoes were held in a bench vise, and the emery cloth was drawn over the edge of the first specimen in a single pass similar to the way a bow is drawn across the strings of a violin.

The second specimen was treated in the same fashion but exposed to three passes of emery cloth abrasion. The third specimen was abraded only along the edge of the shoe and not the walking surface of the pattern area.

Each of the abraded specimens was subsequently used to produce extrusions in modeling clay, then those markings were also photographed.

10.1.5 Discussion

It would appear that shoes of the same design and produced by the same mold can be expected to produce larger features that can be measured in millimeters and smaller features that are more appropriately measured in microns that correspond well from one specimen to another.

It is also clear because of this experiment that the larger features of a pattern, such as the block and voids caused by elements of a pattern, can be expected to recur with a measure of dependability. The smaller features, however, are subject to alteration by even a small amount of abrasion (see Figure 10.1).

These findings seem to imply the following bad and good news:

1. Striations extruded at a similar angle by shoes from the same mold but without significant wear can be expected to yield no more than similar features.

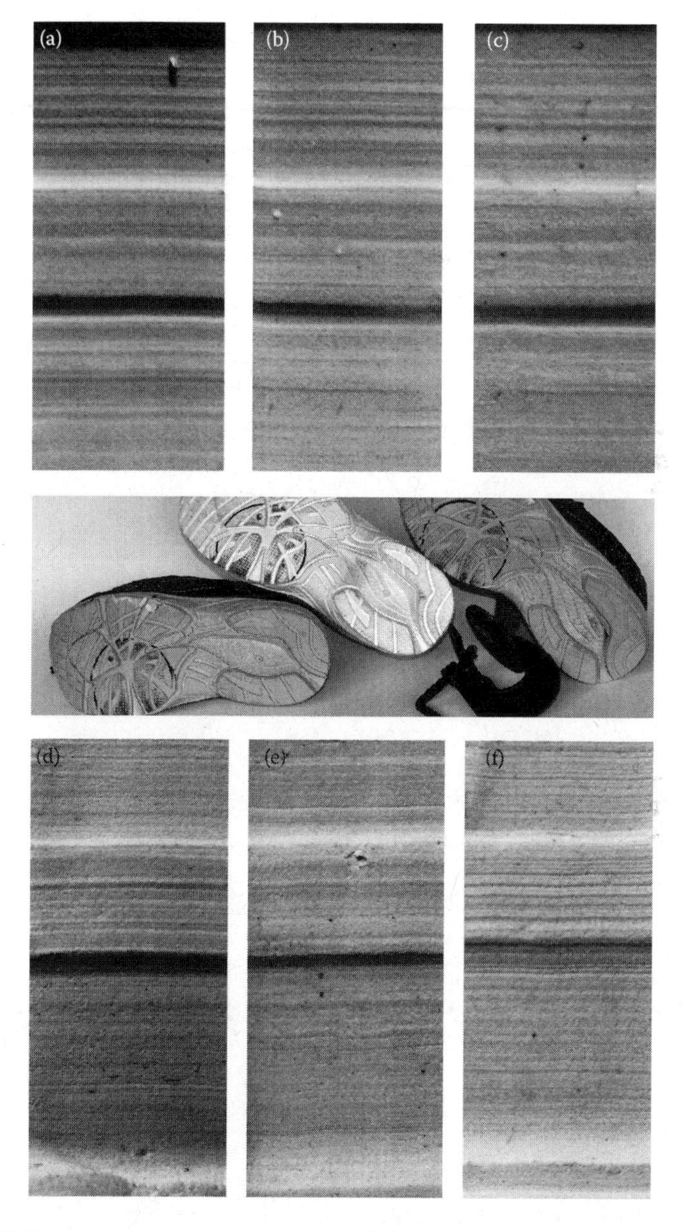

Figure 10.1 Striation studies. Two sets of three impressions are divided by an image of the equipment used to make the markings. The top set shows definitely similar detail from one extruded impression to another. The three impressions at the bottom display considerable differences in many features. The portion in (f) was made with abrasion occurring to only the side of the outsole, while (d) and (e) show a marked change in the width and appearance of both large and small features.

2. Striations that can be attributed to wear between a specimen and the crime scene marking, under the same conditions, may allow some degree of individualization.
3. Outsoles that produce markings that differ from an "off-the-shelf" version must be proven to have originated from the same mold to substantiate any claim of individualization.
4. The correspondence of fine and coarser markings found in the presence of additional random details will allow a relative degree of certainty regarding propositions of common origin.
5. When a time lapse has occurred between finding the crime scene impression and comparison to the suspect shoe, abrasive influences may change many of the finer details.
6. Although steeped in caution, the results of this experiment appear to indicate that differences between the impression and the shoe should be investigated to determine if the suspect's shoe was exposed to some obvious wear or damage.

10.1.6 Edge Characteristics of Footwear

The comparison of striations to a bar code is implied by the voids appearing between elements of a pattern that cause a discontinuous appearance. These bands correspond to the location of pattern features such as the blocks or other features found along the edge of the outsole. Edge characteristics can be made between an outsole positioned at only a few degrees from a parallel orientation to the substrate up to or exceeding a perpendicular inclination of the sole to the substrate.

Where the inclined angle is of sufficient magnitude, characteristics are no longer defined by the sole but often by the adjacent construction of the shoe or boot (i.e., a foxing strip or embossed logo). This condition would be analogous to finding an impression of the sidewall of a tire rather than the tread impression.

Three-dimensional impressions are reproduced with great accuracy in clay-based soils. Many appear in the near-perpendicular range (as made by footfalls in mud). Slippage occurs at varying angles, which can result in less depth in some impressions. A normal step will have a more predictable depth for a given substrate. The ease with which three-dimensional impression angles of inclination can be interpreted or anticipated (by studying additional factors such as gait) does not often apply as easily to two-dimensional versions.

With three-dimensional impressions, the striations can be expected to have rounded, triangular, squared, or irregular features in cross section. The exact formation of these cross sections depends on the shape of the indentations (patterned or randomly generated) on the footwear edge, such as the breaks between blocks of patterns, scratches, or gouges and the amount of penetration of the specimen into the softer substrate. Comparatively, for two-dimensional impressions made by the same areas of the same footwear, the same features, breaks in pattern edges, scratches, and gouges will appear largely as voids, which tend to resemble bar codes even more closely.

Edge features that can be attributed to random characteristics such as gouges, scratches, or cuts may also possess some additional detail because of the extrusion effect on a softer substrate. Markings existing on two or more elements of separately manufactured portions of the shoe, such as those that extend from the sole onto a surrounding foxing strip, can be assumed to have occurred because of an accidental event. The alignment of such markings would not be found on any other specimen in precisely the same manner, if the elements can be shown to have been manufactured and constructed separately.

Care must be taken to ensure that the marking is what it appears to be, and that what appears to be a foxing strip is not a molded feature. One must remember that molded outsoles are popular, and some moldings imitate foxing strips. This type of examination requires much attention, copious notes, and meticulous record keeping.

There are three classifications of striation markings:

Type I: class markings, such as the divisions caused by the patterned elements of the footwear outsole; the stippling found on foxing strips; other such manifestations caused by manufacturing and occasionally by artifacts or accidental damage to a particular mold (Figure 10.2)

Type II: coarse, random markings, such as scratches, gouges, inclusions, and vestiges of damage that can be attributed to a particular source, such as a sharp object at a crime scene

Type III: fine striations

Comparison of two-dimensional and three-dimensional markings created with the same features should then allow determination of the validity of all markings made by the same footwear. Questions remaining after this initial determination process will include the response of different shoes by the same manufacturer. Remaining issues to be explored include the stability of markings under varieties of pressure and angles. The definition of angles and their effects must be isolated and examined (e.g., the influence of roll, pitch, and yaw of the outsole compared to the substrate).

10.1.7 Improvement of Test Equipment

Early studies regarding edge characteristics of footwear[1] were conducted with a homemade device that consisted of a platen suspended on rollers that

Figure 10.2 Examples of naturally occurring striation marks. A form of striated marking that is obviously formed by slippage. The markings were all made following a brief rain shower in two different locations. (a) The soil was part of a landscaped area; (b) the soil for the two markings (made with slightly more pressure on the left impression) was on an exposed bare patch known to have a high clay content. Note that all three markings show some inclination of the soil to exhibit cohesiveness, which causes rupturing of the extrusion in places, and at the top of the impression in right of (b) there is a patch of soil disturbed from which a clod of soil attached itself to the outsole.

allowed horizontal movement of the substrate against a specimen. The device introduced a significant degree of error in both the transport of the substrate and the application of pressure to the clamped specimen tested. Automation of the device was considered a possible solution for improving the accuracy but the platen was found too unwieldy for manual use.

A search of literature on the subject revealed that a British pendulum tester offered a considerably improved method of measuring the coefficient of friction for rubber.[2] This instigated investigation of pendulums for validation studies. Subsequent tests using a pendulum apparatus were much easier to control.

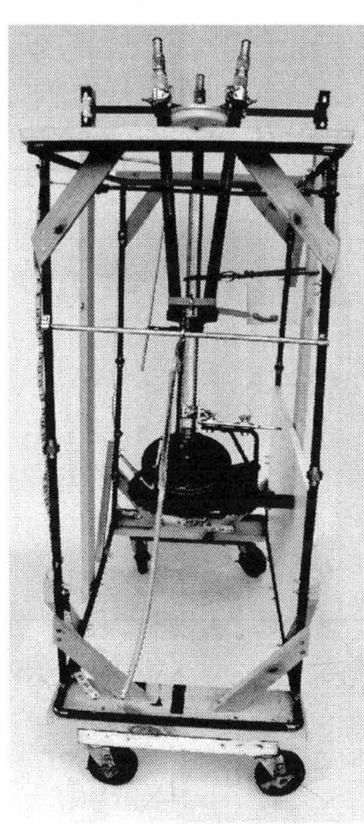

Figure 10.3 Pendulum tester (version 2). The pendulum shown here was constructed to perform three essential functions. (1) The use of a stabilizing wheel allows creation of subsequent manual impressions that are repeatable. (2) The framework provides a reference structure that is used with repositionable scales and lasers to establish and confirm the angle of any adjustments. (3) With the wheel removed, the free-swinging pendulum can be used, in a release-and-catch mode, to allow calculations of the coefficient of friction for various paired surfaces or to examine the effects of high impacts on a single pass.

An earlier model of this device is pictured in Chapter 6. That prototype was used to examine the friction behavior of footwear outsoles. This newer version was designed to be used not only for friction analysis but also for studying the absorption of energy in impact studies, and improved control over the quality of striated test impressions for validation purposes (see Figure 10.3).

Calibration of the improved pendulum apparatus was achieved first using a plumb bob method to align the shaft according to an x axis (taken from the floor) and then was directly confirmed with the crosshair feature of the laser device. Using the laser to measure small increments (centimeters) on the full frame of the device for angle determination afforded greater precision and accuracy in the orientation measurements. The data from measurements were then converted to simple calculations of slope for each adjustment.

The clamp for the specimen was attached to a tripod (included in the construction structure). The tripod controls were used to produce easily altered variations to pitch and yaw. The clamped footwear could be further altered in orientation to offer control over roll. The pendulum (tripod) was fixed to a (bearing-mounted) steel shaft. The frame (copper tubing in this model) and some added staging points (adjustable tape rules) allowed the use of straight edges for alignment or measurement from an (x, y) axis as supplied by a crosshair laser.

Pendulum testers are well accepted as useful test apparatus. A test for hardness of a material under impact stress uses either a V or keyhole notch for standardizing fracture conditions. In those tests, a calculation is derived from measurement of a given weight at a given angle, which produces a lesser angle of follow-through following an impact, thus providing a measurement of the energy absorbed by the impact with the specimen.

This is essentially the same method used to determine the coefficient of friction. In that application, it is the rubber that is expected to yield in response to contact with a surface. To test for the coefficient of friction, the specimen needs to be placed at a specific angle at a specific amount of pressure prior to release at a given angle, which provides the same type of calculation used for impact absorption.

10.1.8 Choice of Substrate

Modeling clay (inexpensive and readily available) was chosen as the standard substrate for all of the three-dimensional impressions. The intention of these studies was to prove the validity of three-dimensional impressions. The conditions under which two-dimensional impressions are created are more complex and require different approaches than included here. Validation of two-dimensional impressions would require an entirely different study.

Modeling clay offers the advantage of retaining detail in a manner that closely resembles field conditions. The quality of detail is acceptable, but there is the matter of controlling the impression depth. This was addressed using a specially adapted molding frame in which cakes or bars of clay could be produced at uniform thickness.

The height of the pendulum was readjusted following each different test situation to ensure that an approximate standard depth was achieved. The results became consistent with practice. A rise in temperature added difficulty to the handling characteristics of the clay. Backing sheets and covers were used to press and roll the clay into the frame and further to simplify its release after molding.

Talcum powder was chosen as a secondary substrate due to the virtual absence of frictional force and ease of reconstituting the test surface. Testing with powder was useful for confirming that the product of the outsole was not uniformly deformed by the substrate. Early results with differing test angles did not show much difference in detail. The lack of change suggested (1) that the details were not as affected by a change as suspected or (2) that the rubber in outsoles might be affected (deformed) during contact (see Figure 10.4).

Figure 10.4 Talcum powder. The problem with talcum powder as a substrate for extrusion relates to the susceptibility of the bulk material to vibration or humidity. This illustration shows some indication of both conditions causing slight interference.

Talcum powder retained a useful amount of detail while offering the advantage of easy preparation for subsequent experiments. Gentle shaking of a set of trays containing talcum powder allows rapid creation of a set of usable surfaces for two different test angles with the same specimen. Following a comparison, one can quickly move on to create another set of tests from the next specimen.

Finding means by which to make the process quicker enables the creation of a larger sample base for statistical purposes. In use, it was found that more attention could be given to the accuracy of the test conditions by streamlining the process. A trade-off was experienced in that the substrate is susceptible to air turbulence and vibration during handling, thus making measurements difficult without using photography.

Testing of both the accuracy and the precision of the apparatus was undertaken by creating subsequent imprints of the same specimen at measured increments of 3°, 5°, and 10°. The angles of incidence remained the same for each specimen tested. The angle variants were labeled as combinations of roll, pitch, and yaw of the specimen in orientation to the substrate.

What follows are descriptions of the test conditions in which the specimen-to-substrate contact was limited to variations of angle. The chosen substrate was pliant modeling clay, which was not expected to generate significant resistance to the relatively solid specimen. The test specimens included new and used outsoles consisting mostly of athletic shoes.

Each pass of the clamped specimen was initially set to cause impression depths of 1, 2, and 3 mm (as measured by a digital caliper known to be accurate to approximately ±0.01 mm). The results gained at varying impression depths affected calculation of the coefficient of friction but demonstrated no change in detail.

10.2 The Hypothesis

10.2.1 Methods of Testing the Hypothesis

The usefulness of ideas must be proven. Much of what is hypothesized in this text requires the corroboration of more experimentation. Hypotheses that withstand the rigors of vigorous scientific testing may attain the merit of having evolved into accepted theory.

Empirical statements derived from scientific experimentation are subsequently subject to evaluation, most often in the form of probabilistic analysis. This is the current standard for evaluating a hypothesis that will eventually be applied to the striation analysis from outsoles.

Mathematical analysis (an alternative to the use of statistical probabilities) seems a simple and reliable process, but it is dependent on both the quality and the quantity of tests. It is a logic-based approach that employs Euler diagrams or truth tables. The drawbacks of the logic system begin with the volume of a model that requires a "sketch" of every possible element and ultimately result in a method that is best suited to simple arguments.

A mathematical logic model could be constructed by application of the following:

1. Assign a letter to each component statement in the argument.
2. Express each premise and conclusion symbolically.
3. Form the symbolic statement of the entire argument by writing the conjunction of all the premises as the antecedent of a conditional statement and the conclusion of the argument as the consequent.
4. Complete the truth table for the conditional statement.

The use of these steps would have only two possible outcomes: If there are no inconsistencies, then the argument is valid; otherwise, it is invalid. If there is any complexity in the argument, there will be a large number of elements and an accompanying greater opportunity for error. These techniques are obviously too unwieldy (in both complexity and the number of questions to be answered) for the purposes of this study, yet they do have some allure as an adjunct to statements of statistical probability. "Instead of discussing the 'probability' of a hypothesis we should try to assess what tests and trials, it has withstood; that is, we should try to assess how far it has been able to prove its fitness to survive by standing up to tests."[3]

A hypothesis will at some point become focused on the answering of an essential question or proposition. To begin this process, it is useful to define the proposition by a series of logical sentences. In the case of establishing the validity of striae, the initial determination is a sentence that describes a single striation mark.

10.2.2 Designing the Test Conditions

The primary implement used to create the test impressions is a pendulum that allows steady movement of the shoe along a chosen path against a given substrate. A platen-based device (an earlier prototype) is used to move the substrate while the shoe is clamped in a given position, and a third device makes use of a rotational component. The devices used here are homemade and somewhat crude.

The pendulum was the most versatile device, well suited to the creation of test imprints and in estimating the coefficient of friction for the outsole products tested. The coefficient of friction for three-dimensional striations has been largely ignored in this study but is considered essential to future tests of two-dimensional striations from the same source (same outsole). The mechanism of testing for the coefficient of friction is to measure the difference of distance traveled by the pendulum following contact with a substrate that incorporates any basic change in the circumstances over a base reading with a particular outsole.[2]

Consideration must also be given to the nature of the materials involved with the creation of striation imprints. Soils and other soft substrate will obviously result in the creation of three-dimensional details, whereas substrate that is harder than the outsole will most often yield two-dimensional results. The careful manipulation of variations in pressure, contact angle, and the presence of contaminants should point to some root causes for both similarities and dissimilarities.

10.3 Empirical Studies

10.3.1 A Study of Angles

Current identification of footwear impressions requires an understanding of the structure of the human foot, manufacturing techniques, class characteristics as compared to individual or accidental characteristics, and the principles and processes associated with identification procedures. It is obvious that the knowledge required for both footwear and tool mark identification would apply equally to edge characteristics.

Protocols for validation were developed that included a sequence of tests including variations of roll (partial rotational movement of the specimen around a longitudinal axis), pitch (toe-to-heel movement), and yaw (orientation in which the specimen was placed horizontally off center or sideways according to the direction of travel). The designated orientations were tested on the pendulum device with each footwear specimen clamped at the appropriate combinations of angles. The initial degree of variation selected was 3° either side of a neutral center point.

Initial research[1] revealed that the edge characteristics derived from footwear outsoles could be expected to display unexpected variations in the recorded striation marks produced. The variations seemed most pronounced with two-dimensional markings, but variants were also observed in the comparison of three-dimensional markings.

Larger features such as the gaps between blocks of the pattern or large gouges were found to reproduce fairly predictably. The product of a gap in the pattern was observed as a rounded, squared, or triangular raised ridge. These attributes of transferred detail suggested that even the smaller features that did not entirely record as well could be identifiable, but their use would require validation.

10.3.2 Methodology

The protocols for creating test impressions required solution of several incidental problems regarding the technique of creating test impressions. The first matter involved the generation of a smooth surface of a reliable thickness with modeling clay. Clay was chosen for its ability to retain detail when reused, providing that it could be re-formed to a consistent thickness.

An adjustable frame was adopted that allowed cakes to be made approximately 10 mm thick and with a desired width (variable), in lengths up to about 200 mm. This frame also provided a means of resetting the height of the pendulum. The thickness of the frame allowed precise changes to be made in the inclination (pitch) as new specimens were loaded for testing.

Protocols for testing each specimen were developed with the intention of allowing narrowing of variables to isolate factors that could account for any changes to the results. It was necessary to establish an expectation of the minimum and maximum angles at which two subsequent impressions began to differ in appearance. This meant a separate series of tests that examined the effect of angles on the separate conditions of roll, pitch, and yaw.

Each specimen was tested as follows:

1. The specimen was secured with two clamps to a specific location along the edge of the outsole at an angle of 80° or less and further adjusted for pitch and yaw.
2. The height of the specimen was adjusted using a portion of the frame that was used as a substitute for the thickness of the clay.
3. The cake of modeling clay was loaded perpendicular to the direction of the motion of the pendulum.
4. The pendulum was forced across the substrate.
5. The cake was relocated to a cutting board where the new impression could be separated from the remaining cake, which could be used for subsequent tests.
6. Each impression was marked to show direction of travel and the combination of angles used in that particular pass of the specimen.
7. The specimen was cleaned of excess clay prior to subsequent use.
8. The specimen was realigned after use to the angles appropriate for the next test configuration.
9. All experiments were conducted between 20°C and 26°C.

The restive state of orientation was designated as 0°. The equation used for each positive (+3°) adjustment was $y = 1/15x + 0$. The equation used for each negative (–3°) adjustment was $y = -1/15x = 0$. The plotted angles were subsequently marked on the frame of the pendulum as scribed by a crosshair laser positioned within 3 m of the target. Subsequent tests were conducted at 5° and 10°.

The choice of using pitch and yaw, without inclusion of the effects of roll, was made based on the observation that roll, in reference to the edge of an outsole, largely results in the use of a different portion of the periphery. It was felt that subsequent tests of different areas should receive separate treatment.

In preliminary tests, ±3° variation of only pitch for a specimen yielded no observable difference in the number or nature of larger and smaller characteristics. The same set of three tests made with ±3° variations of only yaw for the same specimen yielded minor differences in both larger and smaller characteristics.

There were 27 potential combinations of pitch, roll, and yaw in positive, neutral, and negative orientations, all of which were used with the first few specimens to confirm the effects of the separate conditions. This resulted in the generation of many orientations that produced markings with no significant change in detail. In attempting to simplify the process, the following protocol for experimental angle combinations was adopted (see Figure 10.5).

The application of pitch, roll, and yaw to the edge of a footwear outsole does not implicitly depend on the heel or toe of the specimen but only the area of the perimeter examined. Pitch in this usage denotes the forward-and-backward attitude of a segment of the perimeter that could easily be perpendicular

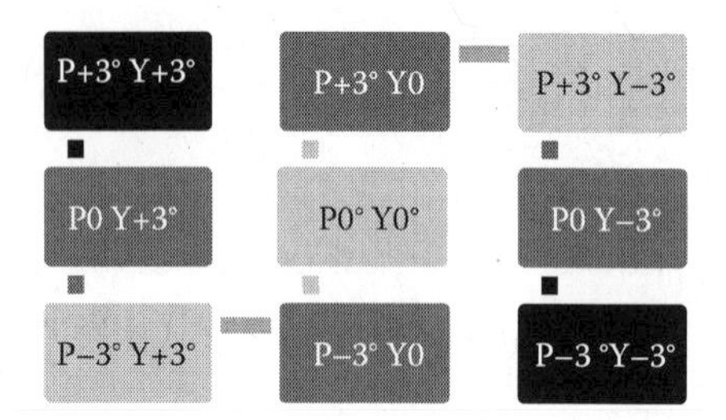

Figure 10.5 Initial testing sequence. The conditions of pitch (P), roll (R), and yaw (R) at 0^0, +3°, or -3^0 of incline are shown in a series of combinations meant to isolate the point at which a change could be predicted.

to the axis of toe and heel. Roll in this respect should be visibly discernible, at least with those outsoles that contained a variety of block elements in the pattern, thus allowing one to readily see just which area of the edge characteristics of the outsole was likely responsible for generating the striations.

Testing of outsoles was eventually pared down to three conditions of a plus, minus, and neutral angle for only pitch and yaw, which could be expanded to the 9 combinations mentioned or all 27 variations if a discrepancy in the nature of a striation was detected that could not be attributed to a single factor (see Figure 10.6).

The lack of difference between variations in pitch was an unexpected result. The same result, however, was observed with subsequent experiments. Another example is shown in Figure 10.7 that tends to support the hypothesis that pitch has a minimal effect on detail.

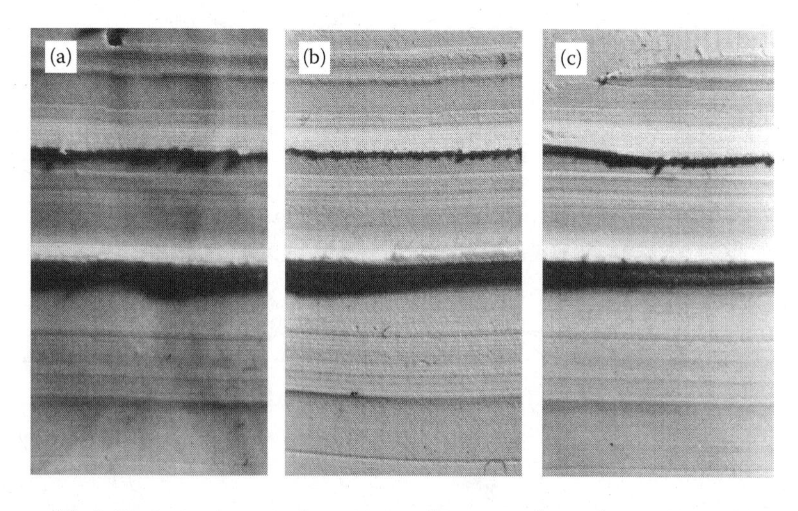

Figure 10.6 Variations in pitch at 3°. Markings made with a well-worn athletic shoe are shown in 3° increments, increasing from (a) to (c). The recorded details show little difference over a 6° difference in pitch.

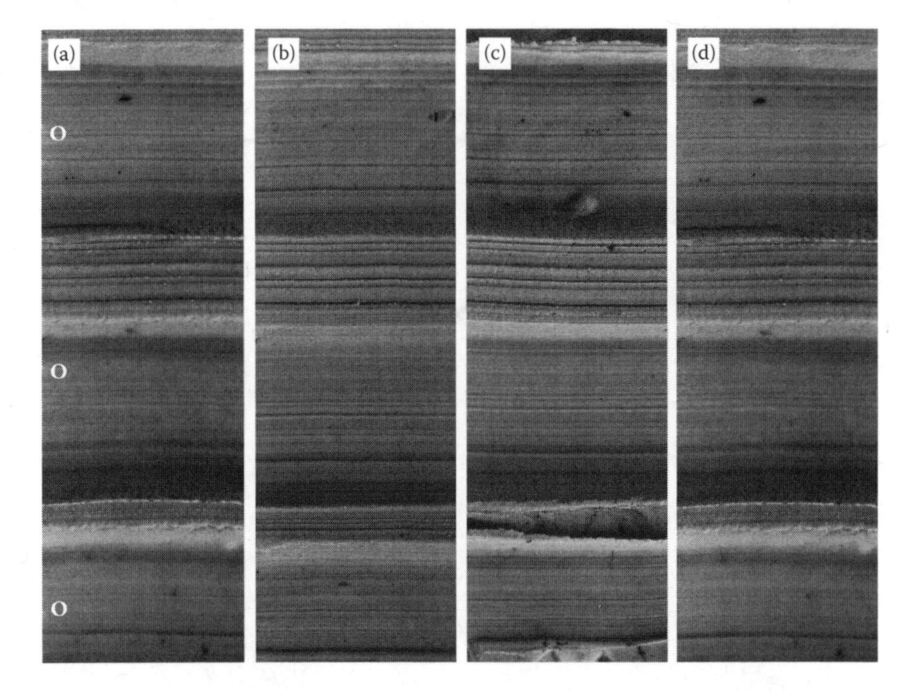

Figure 10.7 Variations in pitch at 10°. Four markings are shown in (a)–(d). The three deepest portions of the markings, designated o along the left, correspond to the top edge of the lugs in the boot outsole. These impressions were created in 10° increments with the edge of a boot outsole to show that the findings correspond fairly well over a total of 30°.

Roll becomes a necessary consideration in the comparison of a crime scene marking to a suspect's outsole. Experiments with the edge characteristics of well over 100 outsoles demonstrated that roll can reliably determine the portion of the edge in contact. It is possible that, with some specimens more than others, roll may have an impact of a synergistic nature, although this was not generally seen as the case. Yaw was the variable that apparently introduced the most change in details, affecting both their relative location and description (see Figures 10.8 and 10.9).

10.3.3 Discussion

This study cannot at this time support a conclusion that striation marks alone are viable sources of individualization. The reason for this statement is that the study is not yet complete. It is apparent from experimentation that the selection of angles can be used in a predictable and measured manner, which means that the analysis of markings can be achieved with some reliability.

The agreement of a substantial number of both fine and coarser features indicates that continued study is needed to create sufficient appropriate

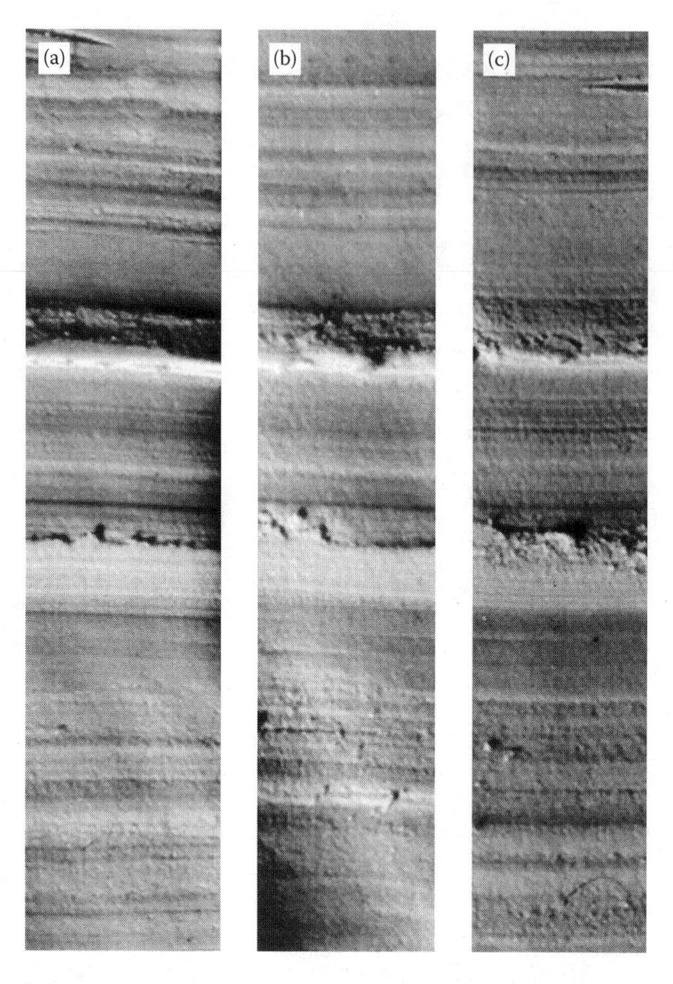

Figure 10.8 Variations in yaw at 3°. With the same athletic shoe outsole used to create the pitch variations in Figure 10.6, the edge of the outsole was changed in 3° increments.

samples. Early studies of the problems also showed the need to formulate a second hypothesis that explains the mechanics of both similarities and dissimilarities, and this appears to have been achieved.

Three-dimensional impressions will generally include notches, voids, or defects that can assist in the selection of an appropriate angle for roll. Roll, without the assistance of random features, must be the first factor to be resolved. Determining the correct angle for roll will leave the angles of pitch and yaw to be established. A uniformly notched edge pattern or entirely smooth edge may require the addition of an exhaustive series of tests to determine the appropriate orientation of the specimen.

Changes in yaw offered the greatest change. One can imagine that a wooden gate held square to the earth will form a set of parallel markings

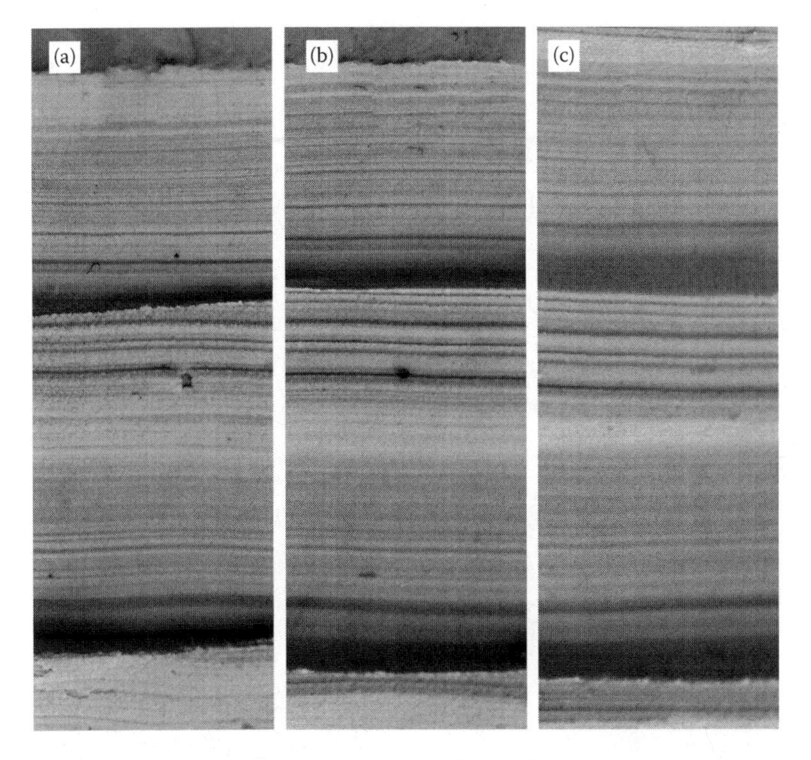

Figure 10.9 Variations in yaw at 10°. Using the same boot outsole that created the pitch variations in Figure 10.7, the edge of the outsole was changed in 10° increments.

that will be narrower if the gate is pivoted to another angle and then dragged in the same direction. Thinking of the edge of an outsole as consisting of a series of gates, we can imagine the effect of yaw in which the locations and appearance of parallel markings can be expected to behave as demonstrated in the aforementioned experiments.

The observations presented here suggest a protocol for creating test impressions. The methods involve

1. Analysis of the crime scene marking
2. Approximation of the roll angle (portion of edge in contact)
3. Approximation of yaw angle (from side to side)
4. Approximation of the pitch angle (tilting forward and back)
5. Fine-tuning of each angle to ensure creation of the best possible test impression for comparison purposes

The work that remains to be done is simply to create a larger sampling. This statement reflects the situation of comparing markings that are not complicated by any other influences such as the dimensional stability of outsoles (a matter that could also benefit from further research).

10.4 Dimensional Stability of Outsoles

10.4.1 Complications Introduced by Dimensional Stability

Variations in the dimensional stability of an outsole, as in the discussion of polymer strength and deformation (Chapter 2, Section 2.4.1) were considered a matter that could potentially affect a worn outsole. Dimensional stability issues within an entire outsole pattern could be difficult to detect, yet the same effects may have much greater implications for edge characteristics. Protocols for testing the effects were devised, and the first experiment is described next.

10.4.2 Equipment

Five representative specimens were selected:

1. A counterfeit shoe (green outsole)
2. A counterfeit shoe (blue outsole)
3. A counterfeit shoe (white outsole)
4. A specimen known to swell with exposure to diesel fuel
5. A shoe identified on the outsole as "oil resistant"

The first three specimens were all new (never worn) counterfeit athletic shoes of the same apparent brand, but different sizes. Specimen 4 was chosen as a control and was an old athletic shoe that had been used in previous experiments; the specimen was known to swell with exposure to diesel fuel. Specimen 5 (a low-cut work shoe) was chosen due to the claim of oil resistance imprinted on the outsole.

10.4.3 Method

Treatment of the specimens included making a test impression in modeling clay as a control impression. A small portion of each specimen (at a location away from the test area) was cut from the outsole and measured. A large covered tank was then filled to a depth of approximately 20 mm with diesel fuel. The cut portions were immersed in the fuel, and the relevant part of the edge of each specimen shoe was allowed to sit in the fuel so that only a relatively small portion of the outsole would be affected.

Exposure times of 144, 192, and 240 hours were chosen. The tests were conducted within the tank as placed in a sheltered location at an average temperature of approximately 29°C daytime and 23°C overnight. The volume of volatile material in the test required sufficient ventilation to prevent any accumulation of fumes.

Specimen 1 was exposed for 144 hours before removal. The cut sample was removed with the shoe, and both were washed in a solution of soapy water to allow limited handling as measurements were taken. The outsole pattern of this specimen was slightly deformed up to about one-half of the width of the outsole (prior to and without visible change after washing).

Specimen 2 was removed after 192 hours of continuous exposure. The cut portion of the immersed specimen had swelled considerably as compared to the area from which it had been cut (not exposed to fuel). The contour of the outsole had visibly swollen on the half that had been exposed (even though the depth of immersion was about 2 mm). These specimens were washed and measured in the same fashion as specimen 1.

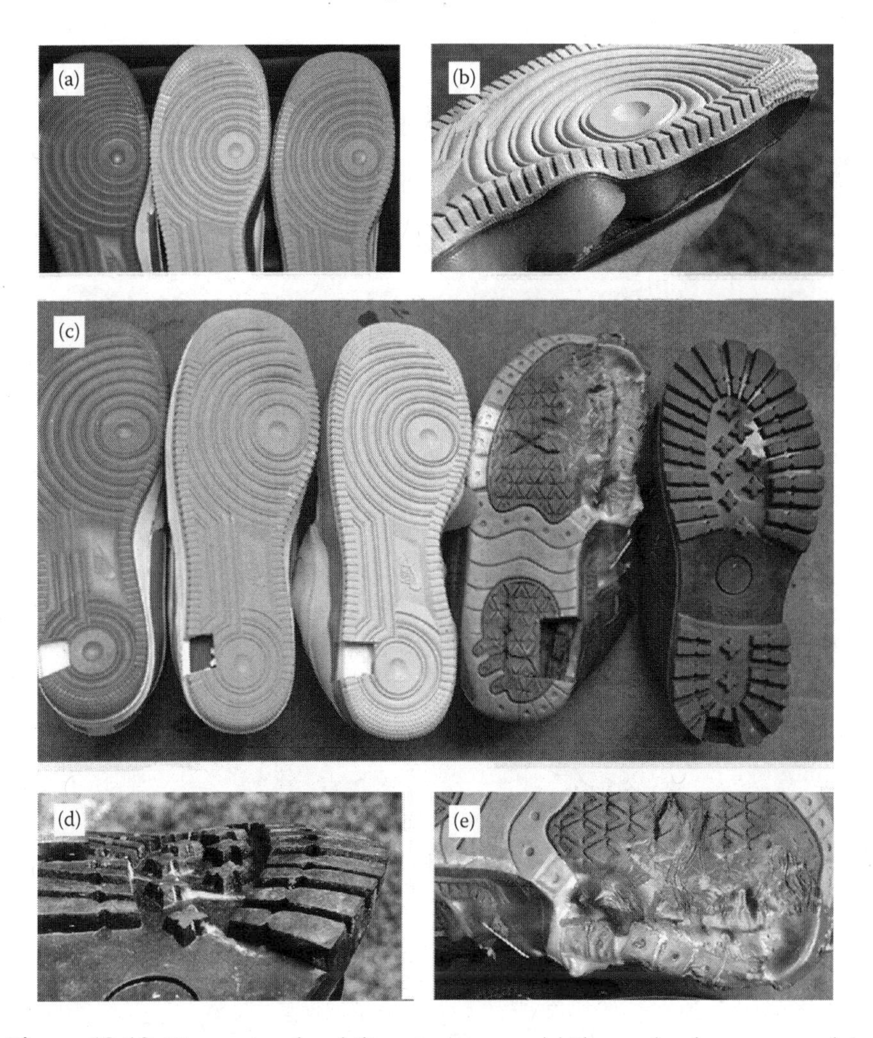

Figure 10.10 Dimensional stability experiment. (a) Shows the three counterfeit specimens before exposure and (b), (c), (d), and (e) show post-exposure results.

The remaining three specimens (and cut samples) were treated somewhat differently in that they were removed from exposure after 240 hours and allowed to drain for 24 hours before measurement. It was noticed that all of the specimens (and cut pieces) were subject to swelling.

Specimen 3 was highly swollen on one-half of the outsole and ruptured along the side of the material adjoining the outsole. Specimen 4 was highly deformed, appeared molten, and actually turned to a liquid state overnight. Specimen 5 was visibly swollen on one-half of the outsole but otherwise was intact (Figure 10.10).

The specimen expected to swell and the counterfeit outsole were exposed to diesel fuel for 6 days (144 hours). At the end of that time, those shoes (and the associated cut samples) were removed from the fuel, cleaned with soapy water, and measured.

At 8 days (192 hours) of exposure, the green outsole and related cut sample were removed, cleaned, and measured. The remaining two shoes and associated cut samples were removed after 10 days (240 hours) of exposure.

10.4.4 Tables of Averaged Values

The tables that follow were created from the response of the cut portions of the outsoles as illustrated in Figure 10.10(c) to immersion in the same diesel fuel as the outsoles they were cut from. The periods of exposure were of the same duration as the outsole from which they were cut. These specimens were briefly washed prior to handling and the measurements were taken out-of-doors as safety precautions. These specimens were measured five times using a digital caliper acurate to ±0.01 mm, and the data in the tables represents an average of those measurements. The tables are arranged to show the measurement prior to exposure and the measurement following exposure for each of the width, length, and thickness of five specimens (Tables 10.1, 10.2, and 10.3, respectively). Table 10.4 shows the calculated percentage of change and Figure 10.11 illustrates those figures in a comparative chart.

This is an anomaly in the findings regarding the fact that the cut section of the outsole that melted (with the same exposure of 240 hours) did not melt. It should be noted, however, that the cut specimen for that outsole was very much affected by the exposure and hard to measure due to a loss of rigidity.

Table 10.1 Values for Width (mm)

Specimen Exposure (h)	1	2	3	4	5
Prior	26.09	23.46	28.02	32.66	30.35
144	29.46				
192		25.63			
240			29.49	36.98	33.73

Table 10.2 Values for Length (mm)

Specimen Exposure (h)	1	2	3	4	5
Prior	27.10	26.06	23.52	49.58	0.843
144	31.59				
192		31.01			
240			29.49	60.47	0.969

Table 10.3 Values for Thickness (mm)

Specimen Exposure (h)	1	2	3	4	5
Prior	12.83	15.18	12.39	14.25	12.04
144	17.02				
192		25.45			
240			13.54	15.72	12.52

Table 10.4 Percentage of Change

Specimen	1	2	3	4	5
Thickness	+32.66	+67.65	+9.28	+10.31	+3.98
Width	+12.92	+9.24	+5.24	+13.23	+11.13
Length	+16.57	+18.99	+25.38	+21.96	+14.95

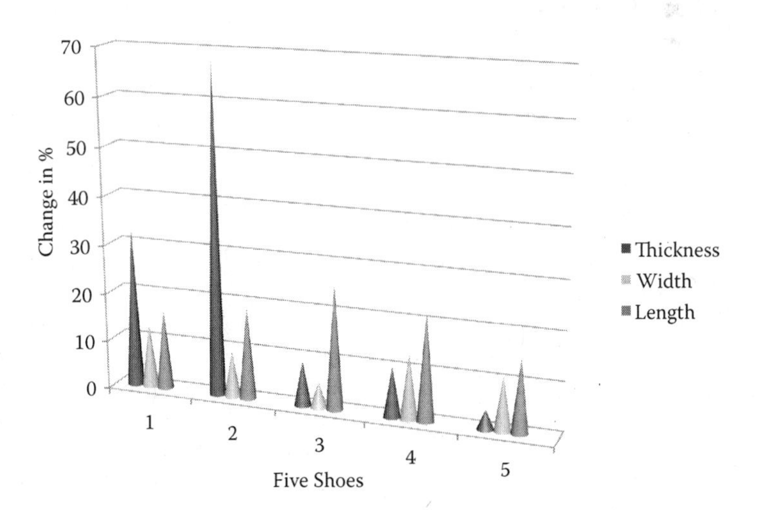

Figure 10.11 Comparison of the change to footwear outsoles exposed to diesel fuel for varying time periods to diesel fuel reveals some interesting trends. Specimens 3, 4, and 5 are the counterfeit soles, all of which show greater swelling in the length of the specimen. The oil resistant specimen swelled more in thickness than length as did the outer sole.

10.4.5 Discussion

Studies of the dimensional stability were a by-product of the recognition that the edge characteristics of outsoles could be identifiable. The tolerance for dissimilarity is much smaller for striation marks than it is for footwear outsoles, where the entire pattern is available for comparison.

For manufacturers of footwear, dimensional stability is a known concern as some outsoles are meant to withstand exposure to harsh conditions. Engineers who have researched such matters are generally concerned with the overall longevity of the product. There was no literature that investigated the potential dimensional stability in a manner that would parallel the issues that could affect a footwear comparison.

The experiment conducted with five shoes pertained only to a continuous exposure of varied duration. The specimens cut from each outsole were entirely immersed, while the remaining shoes were arranged to contain exposures to only the edge of a specific portion of the solvent.

The experiment involving five specimens provided the following observations:

1. Swelling occurred with each specimen.
2. The amount of swelling appeared to be proportionate to the amount and duration of immersion.
3. The response of the counterfeit outsoles seemed consistent.
4. The melting response of the well-aged specimen (known to swell) was unexpected and possibly attributable to warmer ambient temperatures than were present in other tests.
5. The oil-resistant specimen showed the least-damaging swelling effects.
6. Swelling occurred only in those areas that were placed in direct contact with the chosen solvent (diesel fuel).
7. The specimens that could be used to create impressions following exposure were soft and did not yield consistent results under current test conditions.
8. Three new specimens reacted similarly and the other two differed greatly.

The change to the method of measurement (leaving the specimen to drain for 24 hours) caused an interesting phenomenon. Further testing would be needed to corroborate the findings, but draining over an interval appeared to affect the degree of swelling in a specimen. The change of state seen with portions of the specimen 4 outsole closest to the exposed area was unexpected and may be related to the higher ambient temperature during exposure (as compared to previous tests).

The inclusion of an oil-resistant product verified that there was some validity to the claimed resistance. It should be noted that specimens with softer texture during handling included the oil-resistant specimen.

10.5 Experiments with Casual Exposures to Solvents

10.5.1 Periodic Exposure to Solvents

10.5.1.1 Equipment

- Three different shoes (one known to be susceptible to swelling) were chosen for this simple experiment.
- A block of wood was screwed to each of the outsoles that acted as a guide for quickly creating subsequent test impressions in modeling clay at approximately the same setting.
- Diesel fuel was used as the solvent and was applied with a foam brush.

10.5.1.2 Method

Each shoe was basted in a premarked 30 by 60 mm area of the outsole three times daily. Each specimen was allowed to drain and dry between exposures. The treated areas were used to create daily impressions, which were then compared to a control sample taken prior to any application of solvent.

10.5.1.3 Discussion

The results were surprisingly unremarkable. Over a 10-day period of three bastings (a total of 30 exposures) at an average daily temperature of about 20°C to 28°C, no significant variance was observed in any of the specimens.

10.5.1.4 Conclusion

Brief casual exposures to solvents as modeled in this test did not seem to have any significant impact on the dimensional stability of outsoles. This in effect means that the possibility of an observed difference in size (either overall or in only a portion of the outsole) would need to be investigated for a significant exposure to solvents for a prolonged period of time. Conversely, a search for, or examination of, a suspect shoe that is known to sustain prolonged exposures (as in a work environment) can include an analysis that considers the possibility of dimensional stability issues.

10.5.2 Casual Exposure Measured by Weight

Experiments thus far conducted regarding dimensional stability were based on dimensional measurement. The following simple experiment was undertaken to measure the effect of a single exposure to the weight of a specimen.

10.5.2.1 Equipment

- Four athletic shoes of different brands but similar construction, all in new (unused) condition
- Diesel fuel, canola oil, linseed oil, and tap water as solvents in 250-mL quantities
- Four 200 by 600 mm trays that allowed partial immersion of each outsole
- One digital weigh scale accurate to ±0.1 g

10.5.2.2 Method

The four trays were placed on an inclined (about 5°) surface, causing the solvents to create a pool at one end of the tray. The four shoe specimens were weighed prior to placing them in their respective trays to ensure that the toe area of the outsole was at least partially immersed.

The specimens were removed from the trays, allowed to drain for a 10-minute period, and then reweighed. The specimens were set on their side overnight and weighed a third time the following day.

10.5.2.3 Results

There were no significant changes to report in any of the three specimens exposed to canola oil, linseed oil, or water. The only specimen that showed any appreciable degree of change was the one exposed to diesel fuel, which appeared slightly swollen in the area of contact. Three averaged readings were taken for each weight recorded and were as follows:

- Dry weight 308.3 g
- Drained weight after immersion 445.2 g
- Dry weight following one full day of drying 422.8 g

10.5.2.4 Discussion

These results confirmed the relative volatility of diesel fuel compared to the other solvents used. The resulting difference indicated that diesel fuel appears to evaporate from the exposed specimen over time. A different series of tests would need to be conducted to determine the effects of either long-term or transient exposure. The vulnerability of the specimens to damage, before and after exposure, and the effects on both outsole patterns and striations deserve further study.

10.6 Summary of Possibilities and Probabilities of Validation

There are some points raised by the research that has thus far been conducted. The nature of striation marks has been probed and found to be of a relatively short life span during which abrasion plays a role. The markings are distinct and reproducible. Experiments have shown a definite threshold of distortion in terms of the angle at which a marking is made, yet they are retained over an impressive range of conditions.

Studies of the three-dimensional attributes of edge characteristics of footwear created a reasonable groundwork from which to assume that two-dimensional impressions should also be considered as another viable topic for validation research. Problems associated with that assumption are as follows:

1. Three-dimensional impressions are generally formed in a softer substrate. This factor reflects that the dynamics of a two-dimensional impression (usually) exhibit a reversed role of hardness and softness, which will require even more exhaustive validation protocols.
2. A softer substrate allows the effects of roll, pitch, and yaw to be studied without the complication of movement in the elastic features of the specimen. While a harder substrate seems simple enough to account for, one must also take into account the coefficient of friction for the interfaces in a contact. This means that the study of two-dimensional impressions must include the previous features of roll, pitch, and yaw in combination with the added variations of pressure, friction, and surface condition.
3. The effects of pressure, friction, and surface condition would tend to complicate two-dimensional impressions. They were not included in this study because with a softer substrate the effects would be minimal.
4. A substrate that is more pliable than another can possess inclusions that are readily visible or are observed with increased pressure. These features will seldom cause significant variation in impressions with a polymer that is significantly harder than the substrate (obviously, some variations of this arrangement of relative hardness could need specific interpretation).
5. The presence, absence, or decay of a secondary matrix (one of the primary surface-interfering conditions) between the outsole and substrate can be expected to greatly affect the appearance of the crime scene impression and the difficulty of producing a suitable test impression for comparison.

Given the fact that more test impressions will be required to achieve a significant statement of probability, the current evaluation of evidential value for three-dimensional striated impressions is that an identification would require establishing a series of decidedly obvious random contributions. This status may be upgraded with the completion of a sufficient sample base. The current target for future testing is a minimum of 500 experiments, which are hoped to yield a degree of probability in the region of 95% certainty.

The purpose of a validation study would be defeated if the parameters of the data could not be applied to common use. The production of test impressions that match each other quite nicely was not the goal of this line of experimentation, neither in regard to the striations produced nor the effects seen to stem from exposure to solvents or other influences. The achievement here is in the design of parameters that can be applied such as developing an image of known deviations that occur within a small variation of the target value. Knowing how much variation in striation matching can be expected when the text impression is within a few degrees of the crime scene mark is then hoped to provide some numerically acceptable range in which a statistical model will apply.

The amount of work remaining to achieve a complete study is of secondary importance to the development of methods that produce useful results. There is a little more work to be done on the handling and use of the test substrate, refinements to test procedures, obtaining sufficient outsoles (nearly three hundred have been collected), generating the empirical data, and publishing the final results. Only when once published and accepted will this information be of use in case work.

References

1. Pierce, D. S. 2009. Edge characteristics of footwear. *Identification Canada*, 32(1), 4–22.
2. Smith, R. H.. 2008. *Friction of rubber.* CRC Press, Boca Raton, FL.
3. Popper, K. 1961. *The logic of scientific discovery.* Routledge, Boca Raton, FL.

Potential of Electrostatics

11

11.1 Images of Electrons

11.1.1 The Potential of Electrostatic Technology

Separation of a marking from some obscuring influence has long preoccupied practitioners faced with processing a daunting scene or situation. Take, for instance, the matter of friction ridge or footwear imprints that can be identified but not attributed to the event in question. There is frustration attached to the inability to discern the age of a marking.

The sentiments are heartfelt by anyone who has invested a great deal of time attempting to locate "good" evidence only to be defeated by the age question. The proposed electrostatic technique appears to allow differentiation between some items and areas that have been disturbed, within a particular threshold. The concept, although not to be considered evidence in its own right, offers considerable potential for use as an investigative aid.

The technology involves noncontact scanning of appropriate surfaces to locate electrostatic signals. This is the same phenomenon that manufacturers seek to eliminate, a localized buildup of surface charges. Detection of the residual changes in an electrostatic surface charge may be harnessed to track those areas contacted by an outsole, hand, clothing, or even a glove.

The potential for abuse, unfortunately, is almost as great as the promise of benefits. It is hoped that patient research applied to the subject will reveal precisely how useful this information will be in the quest to test the age of a marking. When a reading suggests the occurrence of contact, there is no guarantee that the finding will yield a useful marking.

11.1.2 How It Works

Everyone reading this book is comfortable with the knowledge that a magnetized needle can be used as it aligns with the magnetic field of the earth to indicate the direction of the magnetic pole. A long, thin wire of soft iron that has not been magnetized will also perform like a compass, swinging into line with magnetic fields. The magnetic fields of the earth make a useful starting point in understanding electrostatic fields since both are related.

Exposure to the electric field generated by current running through a wire influences the operation of a compass. This is one of the electromagnetic phenomena described by Maxwell's equations.[1]

James Clerk Maxwell (1831–1879) assembled the laws of electromagnetism into a form of equations. The exact equation is saved for a more in-depth study, but the description here of his work unifies most of the concepts that are central to a basic understanding of the subject.

The first equation describes the role of charges in an electric field as defined by *Gauss's law for electricity.* Observations that like charges repel and unlike charges attract with an amount of force inverse to the square of their proximity is, in this equation, further accompanied by noting that the charge on an insulated conductor moves to its outer surface.

The second of Maxwell's equations described *Gauss's law for magnetism* in that it has not yet been possible to verify the existence of a magnetic monopole. By naming that which could not be found, this equation assists with our understanding of the polar nature of magnetism.

Maxwell's third equation explains the electrical effect of a changing magnetic field as contained in *Faraday's law of induction*, illustrated by creating a current in a loop of wire by placing a bar magnet in the loop.

The fourth equation relates to the magnetic effect of a changing electric field as given by *Ampère's law (extended by Maxwell).* This equation notes that a current passing through a wire will set up a magnetic field near the wire and further explains that the speed of light can be calculated from purely electromagnetic measurements.

The equations possess elegance and serve to explain electromagnetic phenomena in a similar fashion as Newton's laws of motion and gravitation illustrate the basics of mechanics. The scope of the equations includes the fundamental principles of electromagnetic and optic equipment from motors and microwave radar, to microscopes, and even to lasers.

The terms for charges, positive and negative, are attributed to Benjamin Franklin. Further concepts of how charges form electric fields, how currents flow, and how insulators (also known as dielectrics) and semiconductors function begin with Maxwell's equations. Couple to those equations the knowledge that, aside from the process of electrolysis, it is the free electrons that move, and this gives us a starting point for a model of surface phenomena.

Electrostatic charges exist in varying strengths on the surface of matter. The surface charge consists of an array of electrons that reflect the elemental or molecular composition of the substance.

The decay of an area of contact is significant as both a tool and a disadvantage. True, over time (dependent on humidity) the field strength weakens. This characteristic may be used to advantage in the determination of the age of a marking. A single marking that is located at the same location as an electrostatic event possesses a calculable life span within the decay limitations of the substrate under the same conditions. Claims at odds with the facts can then be disputed.

The method requires that the device used to image the electrostatic field disturbance also be equipped with some basic metering capability, including indications of relative humidity, temperature, and perhaps a measure of field strength. The next step is to ensure reliability of the calculations by reference to a table of common effects (using the upper limits) or by separate tests of the crime scene substrate on conclusion of other treatments.

Ideal conditions enable completion of a series of tests that would offer a statistically probable time frame describing the contact period. The decay rate suggests a different prospect. It may be possible to track a particular set of footwear. Contact with carpets can result in the generation of a buildup of friction, thereby creating a discernible path that may be followed, either by an operator using a handheld detector or remotely by attaching a detector to a robotic device.

11.1.3 Why It Is Not Evidence

There is potential that one day, with substantial proof of reliability and with attention to the vulnerabilities of the findings, the use of electrostatics could provide a source of evidence. The vulnerabilities of each finding to subjectivity render the technique to a status of nothing more than a potential investigative lead until proven. The essential problems hinge on whether a particular finding could have a defensible use in court.

Variables that would require further research include, but are not limited to, the following:

1. Humidity, which is the predominant variable in both the longevity and quality of any reading. Increased humidity or, as in areas exposed to varying levels of humidity such as a kitchen, bathroom, or locker room, transient humidity levels may either skew readings or prevent any useful results.
2. The composition of paired surfaces, such as a particular footwear outsole composition on a specific type of flooring, could not be reliably differentiated from similar but different pairings.
3. A reading of electrostatic contact that cannot be distinguished regarding authorship or shape. Hence, if there is more than one mark in the vicinity, it is impossible to detect which marking is responsible for the reading.
4. A result that does not yield more detail than a general area of contact and could not be expected to be used to individualize a contact to the source.

5. Contamination and temperature. Both affect results by degrading the strength, quality, and longevity of the response.
6. Lack of a current safeguard to prevent abuse of attributions or to allow verification of the use of the technique or validity of the results.

The vulnerabilities and disadvantages of using such a device are admittedly damning. There are only two reasons to give this idea a second look: It has the potential to offer a method of discrimination between markings based on the relative age of contact; further, it does work in locating a contact point on some surfaces no matter whether the source was a human hand, a wool glove, or a polymer outsole. There is proof that electrostatic phenomena can be detected, but that usage relates to largely industrial concerns that consider electrostatics as a problem to be eliminated.

11.1.4 Potential Applications: Why Bother?

The concept of electrostatics will need considerable testing before any model could attain the status of a tool for crime scene investigation, even if only as an investigative lead. Let us take a look at the possible uses.

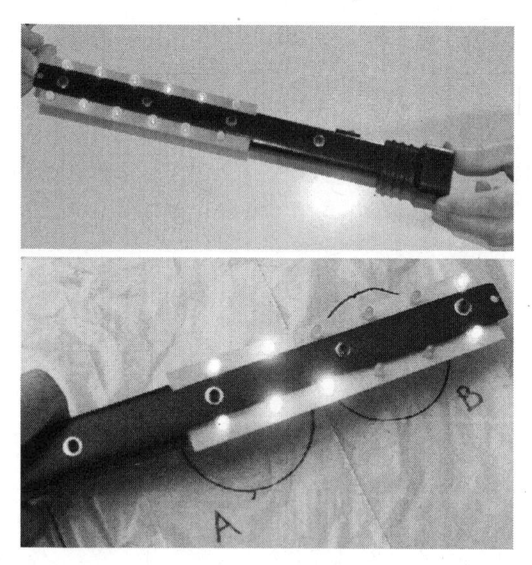

Figure 11.1 Device for detecting electrostatic field disturbances. The prototype wand shown here was designed by the author. The device, used here over two targets marked as circles A and B at about 30 mm above the surface, illustrates that only the thumbprint placed in B causes the lights to switch off. The thumbprint placed in A was made the day before this test, and it no longer causes a response. This wand is far less sensitive than a commercial field tester (also purchased by the author) and does not offer the benefit of a numerical reading, but it does (when calibrated) offer a visual response to markings that have been made within a few hours of scanning.

Tests using a series of sensors arranged in a prototype wand (see Figure 11.1) were capable of discerning electrostatic activity. Used to examine a series of pill bottles, the wand (with a simple on/off response) was able to detect recent activity. In tests in which a woolen glove had contacted a surface, the wand responded to each of the outstretched digits as well as the shape of the palm.

Results of testing with a wand failed to inspire interest in this technology even though it seems to show promise. With further development, this tool could provide a clue regarding which container was recently handled. In an emergency in which a medication or other household poison could have been ingested, the value is without question. It is not proposed as evidence, but it is possible that it could be used to describe the age of the most recent contact, provided that the output is based on proven associations.

What I propose here is to open our thinking to a possible new technique, one that could become a significant tool for at least some cases. The following discussion should provide the forensic community with some reason to consider the usefulness of electrostatic fields in crime scene examinations.

11.1.5 The Triboelectric Series*

When two different materials are pressed or rubbed together, the surface of one material will generally steal some electrons from the surface of the other material. The material that steals electrons has the stronger affinity for negative charge of the two materials, and that surface will be negatively charged after the materials are separated. (Of course, the other material will have an equal amount of positive charge.) If various insulating materials are pressed or rubbed together and then the amount and polarity of the charge on each surface are separately measured, a reproducible pattern emerges. For insulators, Table 11.1 can be used to predict which will become positive versus negative and how strong the effect will be.

Table 11.1 can be used to select materials that will minimize static charging. For example, if uncoated paper (with a positive charge affinity value of +10 nC/J) is squeezed by a pinch roller made of butyl rubber (@ −135 nC/J), there will be about 145 pC of charge transfer per joule of energy (associated with pinch and friction). This is about 20 times more than 7 nC/J, which is the static charge per joule that results from squeezing paper with a roller made of nitrile rubber (@ +3 nC/J). In general, materials with an affinity near zero (e.g., cotton, nitrile rubber, polycarbonate, ABS) will not charge much when rubbed against metals or against each other. The table can also be used (with other formulas) to predict the static forces that will arise between surfaces

* The Triboelectric series and table are extracted from the Alpha Labs web site with the permission of Bill Lee (PhD, Physics).

Table 11.1 Triboelectric Table

Insulator Name	Charge Affinity (nC/J)	Charge Acquired If Rubbed with Metal	Notes
Polyurethane foam	+60	+N	All materials are good insulators (>1,000 T ohm cm) unless noted.
Sorbothane	+58	−W	Slightly conductive (120 G ohm cm).
Box sealing tape (BOPP)	+55	+W	Nonsticky side. Becomes more negative if sanded down to the BOPP film.
Hair, oily skin	+45	+N	Skin is conductive; cannot be charged by metal rubbing.
Solid polyurethane, filled	+40	+N	Slightly conductive (8 T ohm cm).
Magnesium fluoride (MgF_2)	+35	+N	Antireflective optical coating.
Nylon, dry skin	+30	+N	Skin is conductive; cannot be charged by metal rubbing.
Machine oil	+29	+N	
Nylatron (nylon filled with MoS_2)	+28	+N	
Glass (soda)	+25	+N	Slightly conductive (depends on humidity).
Paper (uncoated copy)	+10	−W	Most papers and cardboard have similar affinity; slightly conductive.
Wood (pine)	+7	−W	
GE brand Silicone II (hardens in air)	+6	+N	More positive than the other silicone chemistry (see other silicone entry).
Cotton	+5	+N	Slightly conductive (depends on humidity).
Nitrile rubber	+3	−W	
Wool	0	−W	
Polycarbonate	−5	−W	
ABS	−5	−N	
Acrylic (polymethyl methacrylate) and adhesive side of clear carton-sealing and office tape	−10	−N	Several clear tape adhesives have an affinity almost identical to acrylic, even though various compositions are listed.

Table 11.1 (continued) Triboelectric Table

Insulator Name	Charge Affinity (nC/J)	Charge Acquired If Rubbed with Metal	Notes
Epoxy (circuit board)	−32	−N	
Styrene-butadiene rubber (SBR, Buna S)	−35	−N	Sometimes inaccurately called "neoprene" (see neoprene entry).
Solvent-based spray paints	−38	−N	May vary.
PET (Mylar) cloth	−40	−W	
PET (Mylar) solid	−40	+W	
EVA rubber for gaskets, filled	−55	−N	Slightly conductive (10 T ohm cm). Filled rubber will usually conduct.
Gum rubber	−60	−N	Barely conductive (500 T ohm cm).
Hot melt glue	−62	−N	
Polystyrene	−70	−N	
Polyimide	−70	−N	
Silicones (air harden and thermoset, but *not* GE)	−72	−N	
Vinyl, flexible (clear tubing)	−75	−N	
Carton-sealing tape (BOPP), sanded down	−85	−N	Raw surface is very positive (see above), but close to PP when sanded.
Olefins (alkenes): LDPE, HDPE, PP	−90	−N	UHMWPE given as a separate table entry. Against metals, PP is more negative than PE.
Cellulose nitrate	−93	−N	
Office tape backing (vinyl copolymer?)	−95	−N	
UHMWPE	−95	−N	
Neoprene (polychloroprene, *not* SBR)	−98	−N	Slightly conductive if filled (1.5 T ohm cm).
PVC (rigid vinyl)	−100	−N	
Latex (natural) rubber	−105	−N	
Viton, filled	−117	−N	Slightly conductive (40 T ohm cm).
Epichlorohydrin rubber, filled	−118	−N	Slightly conductive (250 G ohm cm).
Santoprene rubber	−120	−N	
Hypalon rubber, filled	−130	−N	Slightly conductive (30 T ohm cm).

(continued on next page)

Table 11.1 (continued) Triboelectric Table

Insulator Name	Charge Affinity (nC/J)	Charge Acquired If Rubbed with Metal	Notes
Butyl rubber, filled	−135	−N	Conductive (900 M ohm cm). Test was done quickly.
EDPM rubber, filled	−140	−N	Slightly conductive (40 T ohm cm).
Teflon	−190	−N	Surface is fluorine atoms; very electronegative.

Source: Reproduced in whole from Bill Lee (PhD, physics). 2009 by AlphaLab Inc. With permission.

Note: Polyurethane (top) tends to charge positive; Teflon (bottom) charges negative. The charge affinity listings show relative charging. Two materials with almost equal charge affinity tend not to charge each other much even if rubbed together. The third column shows how each material behaves when rubbed against metal, which is much less predictable and repeatable than insulator-to-insulator rubbing. The charging by metal is strongly dependent on the amount of pressure used and sometimes will even reverse polarity. At low pressure (used in this table), it is fairly consistent. N (normal) in this column means the charge affinity against metal was roughly consistent with the value in the second column. W means weaker than expected (i.e., closer to zero than expected or even reversed). The + (plus sign) or – (minus sign) indicates the polarity. In all cases when polarity in the third column disagrees with that in the second column, it is a weak (W) effect.

HDPE, high density polyethylene; LDPE, low density polyethylene; PE, polyethylene; PET, polyethylene terephthalate; UHMWPE, ultra high molecular weight polyethylene; EVA, ethylene vinyl acetate; EDPM, ethylene propylene dienne monomer; BOPP, biaxially oriented polypropylene; PP, polypropylene; PVC, polyvinyl chloride.

and to help select materials that will create an intentional charge on a surface (see Table 11.1).[2]

Limitations of these measurements. Testing was done at low surface-to-surface force (under 1/10 atmosphere) using 1-in. strips of each of the insulators available as smooth solids. (Cotton, for example, could not be made into a solid strip.) The charge affinity ranking of nonsmooth solids was interpolated by their effect on smooth solids that had measured affinity values. At this low surface force (typical of industrial conditions), the absolute ranking of charge affinity of various insulating materials was self-consistent. Above about 1 atmosphere (atm), surface distortions caused some rearrangements in the relative ranking, which are not recorded here. Conductor-to-insulator tests were done also, and contrary to prevailing literature, all conductors had about the same charge affinity. However, the metal-insulator charge transfer was strongly dependent on the metal surface texture in a way not seen with insulator-insulator charge transfer. Metal-insulator transfer was also more pressure dependent in an unpredictable way, so charge transfer has not been quantified for metal-insulator transfer. The "zero" level in this table was arbitrarily chosen as the average conductor charge affinity. "Slow conductors," like paper, glass, or some types of carbon-doped rubber, had approximately the same affinity as conductors if rubbing was done slowly. All tests were done fast enough to avoid this effect. Testing was at approximately 72°F, 35% relative humidity, using an AlphLab Surface DC Voltmeter SVM2 and an Exair 7006 AC ion source to neutralize samples between tests. Resistivities were measured with an AlphaLab HR2 meter. Applied frictional energy per area was 1 mJ/cm^2. Total charge transferred was kept in the linear range, well below spark potential, and was proportional to applied frictional energy per area. All samples needed to be sanded or scraped clean before testing; any thin layer of grease or oil (organic or synthetic) was generally highly positive and would thus distort the values.

Explanation of units "nC/J" used in Table 11.1 (most readers can ignore this paragraph). The units shown in the table are nC (nanocoulombs or nanoamperes per second) of transferred charge per J (joule or watt second) of friction energy applied between the surfaces. The friction energy was applied by rubbing two surfaces together; however, "adhesion energy" might be substituted for friction energy when using the table. For example, when adhesive tape is removed from a roll, a certain amount of energy per square centimeter (of tape removed) must be expended to separate the adhesive from the backing material. Although not yet fully verified, newly dispensed tape becomes charged approximately as predicted by the table if the adhesion energy is substituted for friction energy. After verifying that charge transferred was approximately proportional to the frictional force (for a given pull length), the contact force was adjusted for each pair so that the friction force was 25 g on samples 2.5 cm wide. This is 1 millijoule (mJ) per square centimeter. When a Teflon sample (–190 nC/J) was rubbed in this way against nylon (+30 nC/J), the nylon acquired a positive charge and the Teflon a negative one. The amount of transferred charge can be found by first subtracting the two table entries: 30 nC/J – [–190 nC/J] = 220 nC/J. In this case, using 1 mJ (0.001 J) of friction energy per square centimeter, the charge transferred per square centimeter was 220 nC/J × 0.001 J = 0.22 nC.

"Saturation" or maximum charge that can be transferred. Beyond a certain amount of charge transferred, additional friction energy (rubbing) does not produce any additional charging. Apparently, two effects limit the amount of charge per area that can be transferred. If the spark E-field (10 KV/cm) is exceeded, the two surfaces will spark to each other (after being separated from each other by at least about 1 mm), reducing the charge transferred below 10 KV/cm. This maximum charge per area is about Q/A = 1 nC/cm^2 from this formula. A second, lower charging limit seems to apply to surfaces with an affinity difference of less than (about) 50 nC/J. Two materials that are this close to each other in the triboelectric series never seem to reach a charge difference as high as 2 nC/cm^2 no matter how much they are rubbed together. Although not yet fully verified, it is proposed that the maximum Q/A (in nC/cm^2) is roughly 0.02 times the difference in affinities (in nJ/C) if the two materials are within 50 nC/J of each other. Surfaces that cannot reach spark potential obviously cannot spontaneously dump charge into the air. This is therefore a good reason to select contacting materials such that their affinity difference is small.

Inaccurate information about air being "positive" and the like. A triboelectric series table has been circulating on the Internet, and it contains various inaccuracies. Although attribution is rarely given, it appears to be mostly from a 1987 book. It lists air as the most positive of all materials, polyurethane as highly negative, and various metals as positive or negative, apparently based on their known chemical electron affinities rather than on electrostatic experiments. (From actual tests, there is little or no measurable difference in charge affinity between different types of metal, possibly because the fast motion of conduction electrons cancels such differences.) In gaseous form, air is generally unable to impart any charge to or from solids, even at high pressure or speed. If chilled to a solid or liquid, air is expected to be slightly negative, not positive. There are three cases for which air can charge matter (in the absence of external high voltage). (1) If contaminated by dust, high-speed air can charge surfaces, but this charge comes from contact with the dust, not the air. The charge polarity depends on the type of dust. (2) If air is blown across a wet surface, negative ions are formed due to the evaporation of water. In this case, the wet surface charges positive, so the air becomes negative. (3) If air is hot (above about 1,000°C), it begins emitting ions (both positive and negative). This is thermal in nature, not triboelectric.

References

1. Halliday, D., and R. Resnick. 1978. *Physics*. 3rd ed. Wiley, New York.
2. Lee, B. Triboelectric table. Alpha Lab. http://www.trifield.com/content/triboelectric-series/ (accessed June 26, 2010).

Toward Development of a Unified Theory

12

12.1 Collective Value

12.1.1 Formation of a Theory

According to Ernest Nagel,[1] a theory is formed by three components: an abstract calculus, rules that govern the relation of observation and experiment to the abstract calculus, and the interpretive modeling that allows the concepts to be visualized. Based on this description, a unified theory regarding the mechanics of forensic impressions seems plausible.

The construction of the proposed theory would rely heavily on the familiar hypotheses that are common to the studies of material science and contact mechanics. Formal proof would need to be offered in support of the idea, but experimentation suggests that theories regarding elasticity, friction, and the behavior of materials would provide the source of abstract calculation on which to base a usable theory. The coefficients of friction for rubber and, separately, for metal both apply to many forms of impression evidence. They assist with interpretation of the changing role of soft material (like an elastomer) as it acts on an even softer material or as it is affected by a harder substrate. This also includes the reaction of harder materials (like two metals) as they interact with one another.

Further matters concerning the displacement of materials under load, bulk characteristics, and the deformations that are routinely encountered at crime scenes all seem eligible for inclusion in the proposed theory. Theories of the sort suggested afford a basis for evaluation of subsequent experimentation. The goal of this proposition is improvement in the collective value of observations, experimentation, and testimony.

12.1.2 Diagnostics

Diagnostics are important applications of the knowledge gained from generating models of conditions. Careful use of bench notes will provide the basis of diagnosis of evidence in casework. In disclosure, observations from notes can be compared to known models, which should simplify the delivery and acceptance of testimony.

The tonality of markings, for instance, can be held suspect on finding any one of the following clues:

- the ridge count of the crime scene friction ridge imprint that differs by one ridge
- the presence of dried rings
- the ridges or elements of an impression appear to be lighter than the background
- the marking in question appears as a lighter interruption to background markings (see Figure 12.1)

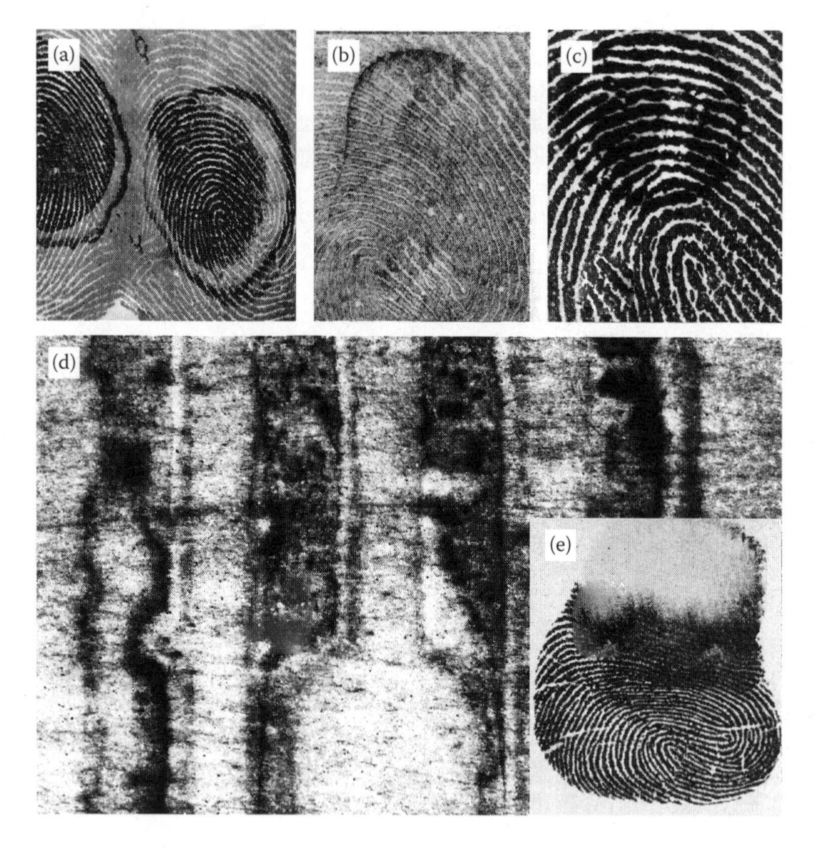

Figure 12.1 Diagnostic clues. All of the images in this illustration were affected by liquids, as evidenced by some degree of capillary drying in each case. The saliva drops on plastic (a), damp raindrop on a car body (b), and symmetric 70% glycol alcohol droplet on plastic (c) were all dried to some degree prior to the application of an overlaid friction ridge imprint on varied substrate. The background impression, a partial damp footwear imprint (d) and the distorted fingerprint (under compressed load from the bottom upward) that is affected by a drop of solvent (e) were both made on paper. Note the tendency of the ridges (b) to demonstrate some capillary effect (darker around the edges of each ridge, especially within the damp area).

Figure 12.2 Diagnostic evaluation. A single wet footwear outsole imprint. It demonstrates a range of features within a single, obviously wet, marking (in this case on a polished metal substrate). Capillary action effects are present, as are the (surface tension-induced) drawing of liquids toward the last point of contact (these marks resemble the veined appearance of surface chemistry markings and some frost marks seen in Figure 9.1).

Mixed-tonality markings are often the most difficult to interpret and are commonly found in areas or on objects that are well handled. A marking that is suspected of mixed tonality presents a good reason for an unseasoned practitioner to seek assistance. The marking will require precise and individual analysis before any conclusions can be proffered (Figure 12.2).

The study of fluids offers a great opportunity to become familiar with a number of characteristic effects. A marking that contains at least some evidence of pooled areas bound by dried rings can be suspected of having been made in the presence of a fluid. The absence of such features may indicate a useful difference between two impressions or conditions.

These interpretations are easily made and can be of evidential value, but their real value will become apparent in more complicated examples for which each practitioner has gained direct experience by conducting his or her own tests. This type of experience will provide a new physical basis with which to describe and chart any findings.

12.1.3 Scope

The domain of evidence can be considered composed of unfathomed depths. Discoveries often begin with a small matter that, on inspection, unfolds to reveal a much larger scope. The sciences have evolved in just this manner, with observations that must be subjected to the larger concepts of corroboration and validation.

All forms of impression evidence are now routinely challenged in court, and that is not necessarily a bad thing. The notoriety has caused much deliberation regarding the methods and practices employed in each case. The attention in turn has caused improvements and an appreciation of the vulnerability of evidence that has not received sufficient consideration.

Several examples of "new approaches" to forensic impression evidence were presented in this book, with at least some indication of the underlying scientific validation. While much of what is examined lacks the finesse of funded science or the benefit of case law, it certainly offers reason to take note of the diverse questions that have not previously been asked or considered.

The path of forensic science is strengthened by criticism. The scrutiny has affected every corner of forensic practice; that in itself is an improvement. There is, however, another guiding method of instituting change that needs contemplation.

12.1.4 Suggesting a Design Path

Some of the matters discussed in this text can be considered informative, while others attack fundamental concepts. Suggestions that point the way to a more unified approach are, however, a catalyst for change. Flaws have been detected in commonly used methods and practices that will not be corrected by the application of superficial or temporary solutions.

Consider that most of the advancements in forensic technology or practices stem from serendipitous accidents rather than as the product of a body of research. The material in this text relies heavily on borrowed technology, and that is not altogether bad, but it cannot be expected to constitute a good method of designing the future path of development.

Scientists develop protocols for their experiments; corporations develop and adopt business plans. These examples point to a need for the forensic community to address the direction of the science by formulating a clear design for the future. A design approach begins with the identification of a need, and the need in forensic science above all else is to maximize or extend the opportunity to locate and make use of evidence. Studying the evidence will lead to an understanding of impression mechanics, which in turn can be used to construct a new unified path.

There can be no doubt that AFIS (Automated Fingerprint Identification System) is an adaptation of a need to fit a technology. The technique is two dimensional (an intrinsic problem), and as pointed out with Figure 2.6, there is a great deal of available data that simply do not appear in current digital impressions. The widespread use of this technology is one of the greatest forensic advancements of the last few decades, but it could easily be thought of as outdated and is due for change (much more change than the addition of a search capacity for palm prints).

A designer, given the task of developing a future system for searching widely diverse friction ridge impressions, would not design a new system that resembles our current technology. Design processes consider all relevant properties, and to design a system that does not reflect or record the anatomy of the subject is a less-than-adequate solution. A less-than-representative design may inhibit the appropriate growth of future technologies, such as three-dimensional modeling techniques (for instance).

12.1.5 Modeling Techniques

The use of modeling techniques to assist comprehension when describing matters of some complexity is a well established practice. Models can rely on any of the senses and often depend on some common experience or shared logic. Literary use of models includes analogies and parables, while mathematical examples can be typified by the use of an abacus, or the many types of numerical systems that permit calculation from image-based notations such as cuneiform, to the binary code of computer language. Successful use of descriptive models introduces clarity while rendering complicated theories more palatable, their use has paved the way for many things such as navigation of the globe.

The inspiration for a particular model stems from experience, and the effectiveness of that model will hinge on the ability of the model to provide a connective association between known or easily accepted quantities or entities and the matter under consideration. The study of DNA from wood fibers is a model scenario (Chapter 1) that could have been illustrated by a somewhat different example. A particular case, encountered before plant DNA analysis was available, involved the collection of a short thorn from the sidewall of a tire. Combined, the species type, rarity of the thorn, coupled with conventional tread evidence; and soya beans lodged in the vehicle underbody would have been sufficient evidence to support, with a high degree of probability, a hypothesis that the suspect vehicle had been present at the scene in question. While the use of plant DNA is not impression evidence, such models can greatly enhance the value of the associated evidence. These two examples, thorns and pry mark fibers, can be expected to have relevance only as a rarity in case work. Extension of the model, however, may lead one to consider sources such as the collection of materials on a radiator or the wheel wells of a vehicle, both of which can trap substantial amounts of pollen, insects, or soil samples as corroborative evidence.

We cannot know with any great amount of certainty what constituents are present in an impression. We may however, by the use of models, make observation of consistencies and inconsistencies. Knowledge of a specific material's behavior can stem from modeling in which the effects can be applied to case work. Deformations, symmetries, elasticity, conductivity, are a few of many observable traits of a physical or chemical nature that can bolster evidence where observations corroborate facts that either support or refute a given hypothesis. The growth of interest in learning about such diverse matters can be expected to provide the impetus to develop a model that encompasses the effects of these and other situations.

Fingerprints, as they are commonly recognized or used, are modeled as two-dimensional images, which do not provide representation of the friction skin under varieties of stress. Technology exists in the form of point cloud surface rendering that, in future, may be adapted to create a three-dimensional model of the entire hand. Such technology, combined with the opportunity to model the effects of load on the digit would achieve two significant effects; firstly the ability to replicate distortion under specific conditions thus providing an improved means of understanding, evaluating, or presenting the evidence, and secondly, a means of better understanding the effects of a matrix on a particular substrate as well as the role of each element of the structure of the digit or palm extending, not just around the circumference, but also including the tips, proximal phalange, wrist, and back of the hand. It is possible to imagine a future system wherein, like today's photographic editing software, one could, with the application of a filter or two, create a directional force on a curved surface, change that to a flat or irregular surface and layer in effects such as the presence of a raindrop or a film of some other lubricant. This type of model should be considered as a potential direction for the advancement of impression technology, applicable to friction ridges, footwear, tires, and tool impressions.

The benefit of a unified theory may be seen as mathematical modeling wherein distortion becomes potentially quantifiable. To visualize this concept, imagine that a footwear impression is plagued firstly by swelling of the elements in two areas, secondly by differential slippage in one area of the mark, and lastly by capillarity effects (ring like evidence of the marking having been made when wet). Assume that the suspect outsole, seized two days later, is not the same size as either of an off-the-shelf new outsole, nor the crime scene marking, which is considered good news in terms of subsequent analysis. Bench marks noting the above would also include whether the size differences are lesser or greater thus indicating the possibility that the off-the-shelf model is the wrong size or that the outsole has recovered in size in the interval between making the crime scene marking and having been seized. Learning the likely history of the outsole prior to the event and during any interval, possible causes of moisture or incidence of inclined planes at the scene, particular details of the suspect (gait or stance for instance) that would contribute to the "fit" of other data, and obtaining detailed measurements of the outsole pattern should allow a number of precise observations to be made. A history of exposure to solvents coupled with periods of freedom from exposures may explain how the same outsole can appear as two different measurements over a two day span and different again in comparison to a new product. The presence and influence of capillary effects, particularly around an area of slippage, can be simulated and compared to the coefficient of friction for dry conditions and the inclined angle of the substrate can be added to the calculations. Remember that all of this data is not inclusive of any identifiable characteristics and that the comprehensive nature of the examination would be enhanced by this type of intensive analysis.

The theoretical scenario given above is not comprehensive in that other relevance can be discovered as mathematical modeling is applied to that which is observed. The same type of analysis can go far toward the examination of any other impression endowing some with greater precision about finer details and perhaps assisting with the determination of the randomness of what could be a manufactured characteristic. Friction ridge markings possess greater flexibility than most outsoles. Measurement of the ridge locations without significant stress and under load can be expected, in both types of donor material, to yield a different coefficient of friction and combinations of measurements could yield some data of use in comparing and explaining the appearance of areas around obviously distorted features. Tire track comparisons can often display minor discrepancies in length during a rotation or even in the sizes of individual elements and tool marks that are large enough, can also exhibit explainable differences, these too would benefit from more precise analysis.

Validation studies provide us with models that have empirical data capable of supporting a statistical evaluation. In consideration of a validation study, we accept that a statistical evaluation is the goal of the process, that the empirical products are the basis of the statistical analysis, and that the protocols and methods of the experimentation or research are vital to developing useful data and subsequent conclusions. In the chapter concerned with the validation of edge characteristics presented here, you will not find reference to a completion of the empirical testing or conclusion, since that testing has been deferred while the protocols were developed. Remember that footwear impressions do not behave like metal, they do not (usually) travel in a helical path similar to a bullet's trajectory, and these as well as considerations of the substrate and matrix involved require a different form of evaluation. In short, it would be easily possible to claim that by reproducing the precise conditions of contact with a three-dimensional surface one can faithfully reproduce the striations from one marking to another, but that would be an approach flawed by the impracticality of accurately reproducing the exact conditions in case work. The approach illustrated about the validation of edge characteristics

(Chapter 10), illustrates a novel method of testing and consideration of variables that allow some latitude in the test result, some expectation of error, and a methodology that should produce useable conclusions.

A manufacturer who understands your needs more fully than his competitor will, predictably, make a more useful product. Creation of a three dimensional impression search engine, or even an electrostatic device that permits some measurement of contact evidence before any other treatment is used are examples of technology waiting to be developed. This, an unprecedented era of change in forensic practice, is the ideal opportunity to shape the direction and needs of forensic theory and practices for the future.

12.2 Cautions

The world is filled with hazards and research of any kind may pose risks that are as easily overlooked as a footwear impression on a piece of fruit or the dimensional stability of a polymer. This text advocates that practitioners within each discipline conduct experiments in order to develop their understanding of how common materials interact. In doing so, the practitioner that conducts experiments must assume responsibility for exposure to such things such as airborne pollutants, the presence of solvents in hand cleaner, or a myriad of other potential health issues.

If your work environment specifies validated procedures for personal protection and equipment use, then the onus is on you to apply due diligence in the use of those guidelines. Where no immediate guidance is available, you can obtain material safety data sheets online or, by calling the manufacturer. There is no excuse for using less than optimal protection when there is a possibility of exposure to toxic materials.

References

1. Nagel, E. 2007. Experimental laws and theories. In Yuri Balashov and Alex Rosenburg (Eds.), *Philosophy of science contemporary readings*. 2007. Routledge, Boca Raton, FL, pp. 132–140.
2. Hildebrand, D. S. 2007. *Footwear, the missed evidence*. Staggs, Wildomar, CA.

Terminal Velocity Calculator

```
/*****************************************************************
Terminal Velocity Calculator Prepared at UWO Research Park
For the book: Mechanics of Impression Evidence
Prepared by: Patrick Mallay
pmallay@uwo.ca
This program asks the user to input the density, and volume of a drop of
liquid and then calculates the height from which it must be dropped to
reach
its terminal velocity.
*****************************************************************/
#include <iostream>
#include <stdio.h>
#include <math.h>
#include <iostream>
#include <iomanip>
using namespace std;

double terminalvelocity(double volume, double density);
double velocity(double terminalv, double vpt2,double x);
double velopt2 (double volume, double density);

int main()
{
//Declare and initialize objects
double volume,density,x(0),y,z,distance, w,
integralpt2,stension ;
char progkill ('y');
//Print question title
  cout <<"*****************************" <<endl
       <<"*                           *" <<endl
       <<"*   UWO Research Park       *" <<endl
       <<"*   Drop Height Calculator  *" <<endl
       <<"*                           *" <<endl
       <<"*****************************" <<endl;
```

```
//Print the purpose of the program
cout <<endl <<"This program asks the user to input the
density, and volume of a drop of " <<endl <<"liquid and
then calculates the height from which it must be dropped
to reach " <<endl <<"its terminal velocity." <<endl<<endl;
//Request the user to enter a number between 10 and 100

do      {
        cout <<"Please enter the surface tension of the
        liquid being tested in dynes per cm" <<endl;
        //Store the input number as surface tension
        cin >> stension;

        //If the number is not between 0 and 500, ask for
        another number until one within the specified range
        is input
        while (stension<0) {
                cout <<"Sorry the number you entered is not
                between 0 and 500" <<endl <<"Please enter a
                number between 0 and 500." <<endl;
                cin >> stension;  }
        while (stension>500)     {
                cout <<"Sorry the number you entered is not
                between 0 and 500" <<endl <<"Please enter a
                number between 0 and 500." <<endl;
                cin >> stension;  }
                cout <<"Please enter the density of the drop
                liquid in kilograms per litre:" <<endl;
                //Store the input number as density
                cin >> density;

    //If the number is not between 0 and 10, ask for another
    number until one within the specified range is input
        while (density<0)  {
                cout <<"Sorry the number you entered is not
                between 0 and 10." <<endl <<"Please enter a
                number between 0 and 10." <<endl;
                cin >> density;          }
        while (density>10)       {
                cout <<"Sorry the number you entered is not
                between 0 and 10." <<endl <<"Please enter a
                number between 0 and 10." <<endl;
                cin >> density;    }

    //Calculate the volume of a drop of the liquid from a
burette with diameter 0.5 mm
volume=(2*3.14159*0.000500*0.001/9.8)*(stension/density);
```

```
//Calculate the terminal velocity of the drop
y= terminalvelocity (volume, density);

   //Determine the constant value for the second half of
   the velocity-time equation
z=velopt2(volume,density);

//Find when the drop reaches its terminal velocity
       do      {
               x= x+0.00001;
               }
       while (((y-0.00255)>(velocity(y,z,x))));

//Calculate velocity at time x (test)
w=velocity(y,z,x);
//Calculate  second part of integral
integralpt2= cosh((z*x));
//Evaluate the integral
distance= (log (integralpt2))*(y/z);
//Display the calculation results
cout<<endl<<"The drop will have a volume of "<<setprecisio
n(2)<<volume<<"L."<<endl;
cout<<endl<<"The terminal velocity of the drop will be:
"<<setprecision(2)<<y<<"m/s." <<endl;
cout<<"It will reach this velocity after travelling "<<set
precision(2)<<distance<<"m in a time of:
"<<setprecision(2)<<x<<"s." <<endl;
cout<<endl<<"Would you like to perfom another calculation?
(y/n)"<<endl;
cin>> progkill;
cout<<endl<<endl<<endl;

}
while (progkill == 'Y' || progkill == 'y');

       //Exit Program
       return (0);
}

       double terminalvelocity(double volume, double
       density)
               {
                       double a;
                       a= sqrt((19.62*volume*density)/(0.15
0796*(sqrt((0.00418879*volume)))));
                       return (a);
               }
```

```
        double velocity(double terminalv, double vpt2,double x)
               {
                         double b;
                         b= terminalv* tanh((vpt2*x));
                         return b;
               }

        double velopt2 (double volume, double density)
               {
                         double c;
                         c= sqrt
((1.479312149*(sqrt((0.00418879*volume)))/
(2*volume*density)));
                         return c;
               }
```

Index

N

O

S

T